VALENCY AND
MOLECULAR STRUCTURE

VALENCY AND
MOLECULAR STRUCTURE

E. CARTMELL, B.Sc., F.R.I.C.

*Lecturer in Inorganic Chemistry
and Deputy Director of Laboratories,
University of Southampton*

and

G.W.A. FOWLES, Ph.D., D.Sc., F.R.I.C.

*Professor of Inorganic Chemistry,
University of Reading*

THIRD EDITION

D. VAN NOSTRAND COMPANY, INC.
PRINCETON, NEW JERSEY

BUTTERWORTHS
LONDON

First Edition, 1956
Second Impression, 1957
Third Impression, 1958
Fourth Impression, 1959
Fifth Impression, 1959
Sixth Impression, 1960
Second Edition, 1961
Second Impression, 1962
Third Impression, 1963
Fourth Impression, 1964
Third Edition, 1966

Printed in England by
ADLARD AND SON LIMITED
London and Dorking

PREFACE TO THE THIRD EDITION

In this new edition we have kept to the aims expressed in our first preface, since there still seems to be a need for a text of this size and at this level. Some alterations to the order in which the material is presented have been made, together with rather more use of molecular-orbital theory in discussing the bonding in simple compounds. Details of structures, bond lengths and bond angles have been brought up to date, and we have again given very full references to the original papers. During the ten years that have elapsed since the first edition of this book appeared we have received much friendly and constructive criticism and we should like to take this opportunity of thanking our correspondents.

E. Cartmell
G. W. A. Fowles

April, 1966

PREFACE TO THE SECOND EDITION

PART I of this new edition is virtually unchanged, but Part II has been re-arranged and revised, with a somewhat greater emphasis on the molecular-orbital method; Part III has been almost completely re-written. The sections on the hydrogen bond and on bonding in metals have been expanded, but the major changes are in the chapter on complex compounds (where we have included an elementary introduction to the 'ligand-field' theory which has had such a stimulating influence on research on compounds of the transitional metals) and in the survey of the structural chemistry of the non-transitional elements.

We are very grateful for much helpful comment by readers of the first edition of this book. We emphasize again that this is an *introductory* text, and to the standard texts referred to in the Preface to the first edition we wish to add W. Kauzmann's *Quantum Chemistry* which we have found particularly helpful. Values of bond lengths and angles have, in general, been taken from the *Table of Interatomic Distances* published by the Chemical Society of London. This invaluable compilation covers the literature up to the end of 1955; we have given full references in discussing structures determined or corrected since this day.

E. C.
G. W. A. F.

The University,
  Southampton
May, 1961

PREFACE TO THE FIRST EDITION

THIS book is written by chemists for chemists. It is mainly intended for first year honours students, although Part III, in which we discuss recent work on the nature of the bonds in many molecules, is suitable for more advanced courses.

Students often dislike systematic inorganic chemistry, largely because of the numerous facts that they must necessarily learn. The subject can be revitalized, however, if the facts are discussed in a setting of structural theory provided, on the one hand by a study of crystal chemistry and, on the other hand, by the application of quantum theory to valency problems. The essential features of crystal chemistry do not require elaborate mathematical formulation for their interpretation, and undergraduates can be referred to several admirable standard texts. It is more difficult, however, to recommend a suitable *introductory* text on the quantum theory of valency. Very elementary, non-mathematical accounts of quantum theory do not usually provide sufficient intellectual satisfaction for the honours student, and the advanced works must necessarily devote a great deal of space to the rigorous treatment of the hydrogen atom before the theory can be applied to valency problems. We have felt the need for an introductory text in which an elementary account of the development of the quantum theory is combined with a survey of its application to structural chemistry. In attempting to provide this at a level suitable for first year students we have necessarily quoted many results without proof; we have tried, however, to explain the reasons underlying the mathematical techniques adopted and the assumptions made, so that the reader may accept the results without feeling that the procedure is quite arbitrary. The application of quantum mechanics to chemistry by non-mathematicians must always be a hazardous enterprise; we hope that we have avoided major mathematical pitfalls.

The experimental foundation of the quantum theory, and the application of the theory to isolated atoms, are described in Part I of the book. Part II begins with an historical outline, tracing the development of the idea that chemical forces are electrical in nature. The two important approximation methods of applying the quantum theory to atoms in combination—valence-bond (resonance) and molecular-orbital methods—are then discussed. The molecular-orbital method is the more fashionable at the time of writing, but a combination of resonance and molecular-orbital language frequently gives the most vivid description of chemical bonds to the (mathematically) unsophisticated student. Both these methods are therefore used in the chapter on directed valency at the end of Part II. Part III contains an account of bonding in solid structures and a survey of recent views on the nature of the bonds in a large number of simple molecules. Bonding in complex compounds is discussed in Chapter 12, and the final chapter is concerned with the interesting valency problems that arise in considering the structures of 'electron-deficient' molecules such as the boron hydrides and some of the metal alkyl compounds.

Our debt to the authors of standard works is considerable; we must make particular mention of C. A. Coulson's *Valence* and L. Pauling and E. B. Wilson's *Introduction to Quantum Mechanics*. It is hoped that beginners, especially those with a limited mathematical background, may find the present book a helpful introduction to such works. In Parts I and II references have only been made to a few papers of historical importance in the development of quantum theory, but much of the information in Part III comes from recent work, and here we have given full references.

We thank Dr. Peter Woodward for his comments on Part I of the book. We are especially grateful to our friend and colleague Dr. K. R. Webb, who read the whole of the manuscript; his critical eye detected many mis-statements and ambiguities. Any remaining errors must, however, be the sole responsibility of the authors. We extend to Joyce Fowles our warm appreciation of her work and care in the preparation of the manuscript, and acknowledge with pleasure the encouragement of Professor N. K. Adam and our colleagues in the Chemistry Department at Southampton.

E. C.
G. W. A. F.

The University,
 Southampton
January, 1956

CONTENTS

II—QUANTUM THEORY OF VALENCY

III—THE APPLICATION OF THE PRINCIPLES OF CHEMICAL BONDING

Periodic Classification of the Elements

IA	IIA	IIIA	IVA	VA	VIA	VIIA	VIII			IB	IIB	IIIB	IVB	VB	VIB	VIIB	O
H 1																	He 2
Li 3	Be 4											B 5	C 6	N 7	O 8	F 9	Ne 10
Na 11	Mg 12											Al 13	Si 14	P 15	S 16	Cl 17	Ar 18
K 19	Ca 20	Sc 21	Ti 22	V 23	Cr 24	Mn 25	Fe 26	Co 27	Ni 28	Cu 29	Zn 30	Ga 31	Ge 32	As 33	Se 34	Br 35	Kr 36
Rb 37	Sr 38	Y 39	Zr 40	Nb 41	Mo 42	Tc 43	Ru 44	Rh 45	Pd 46	Ag 47	Cd 48	In 49	Sn 50	Sb 51	Te 52	I 53	Xe 54
Cs 55	Ba 56	* 57–71	Hf 72	Ta 73	W 74	Re 75	Os 76	Ir 77	Pt 78	Au 79	Hg 80	Tl 81	Pb 82	Bi 83	Po 84	At 85	Rn 86
Fr 87	Ra 88	† 89–															

* Lanthanide Series

La 57	Ce 58	Pr 59	Nd 60	Pm 61	Sm 62	Eu 63	Gd 64	Tb 65	Dy 66	Ho 67	Er 68	Tm 69	Yb 70	Lu 71

† Actinide Series

Ac 89	Th 90	Pa 91	U 92	Np 93	Pu 94	Am 95	Cm 96	Bk 97	Cf 98	Es 99	Fm 100	Md 101	— 102	Lw 103

I

QUANTUM THEORY
AND ATOMIC STRUCTURE

THE EXPERIMENTAL FOUNDATION
OF THE QUANTUM THEORY

In 1900, MAX PLANCK[1] published a theory that explained the results of experiments on the energy radiated by hot bodies. He introduced a new concept into science—the energy quantum—a concept that was applied by EINSTEIN[2], in 1905, to explain the photoelectric effect, and by BOHR[3], in 1913, to explain the spectrum of the hydrogen atom. *The Planck quantum theory is the foundation of modern chemistry.* We shall discuss the work that led to the formulation and general acceptance of the theory in chronological order, although the photoelectric effect and the problem of the structure of the hydrogen atom, are of more immediate interest to chemists.

BLACK-BODY RADIATION

The rate of production of energy by a hot surface depends upon the temperature, the nature and the area; a dull black surface radiates more energy per second than a polished one of the same area and temperature. The blacker the surface the greater the radiation, so that the maximum radiation at a given temperature will be that from a 'perfectly black' surface. The dependence of radiant energy upon temperature alone can therefore be examined if such a radiating surface is available. The radiation is called 'black-body radiation', and although it is impossible to make a 'perfectly black' surface, radiation with characteristics very close to those expected from such a surface can be obtained experimentally from a small opening in the wall of a furnace kept at constant temperature. If radiation from such a source is dispersed by a prism system and allowed to fall on a sensitive energy detector such as a thermocouple, the distribution of the energy among different wavelengths can be studied. The classical experiments in this field were those of LUMMER and PRINGSHEIM at the end of the nineteenth century. A typical result is shown in *Figure 1.1,* where E_λ corresponds to the radiant energy emitted per unit wavelength interval per unit area per second; $E_\lambda d\lambda$ will measure the energy radiated between the wavelengths λ and $\lambda + d\lambda$. It will be seen that only a small amount of the total energy is radiated at very short or at very long wavelengths, and that the curve passes through a maximum.

Two important attempts were made to explain these results. WIEN, using the methods of classical thermodynamics, derived the following expression

$$E_\lambda d\lambda = c_1 \lambda^{-5} e^{-(c_2/\lambda T)} d\lambda \qquad \ldots (1)$$

where λ is the wavelength, T the absolute temperature, and c_1 and c_2 are constants. E passes through a maximum value as λ increases, and with a suitable selection of values for the constants

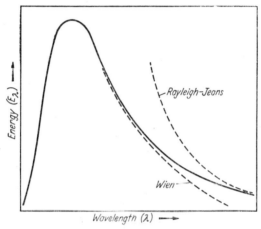

Figure 1.1. Distribution curves for black-body radiation

c_1 and c_2 good agreement with experiment is obtained for small values of λT. At long wavelengths, however, in the infra-red region of the spectrum, the agreement is not satisfactory. The dotted curve in *Figure 1.1* shows the results predicted by the Wien theory compared with the experimental results given by the full line.

The problem was tackled from a different point of view by RAYLEIGH and by JEANS. They worked out the number of 'modes of vibration' of electromagnetic waves in an enclosure, and assumed that the total energy was equally distributed among these modes. This gave the relationship

$$E_\lambda \, d\lambda = \frac{c_1}{c_2} \frac{T}{\lambda^4} d\lambda \qquad \ldots (2)$$

There is good agreement with experiment at long wavelengths

where the Wien equation breaks down, but the equation is unsatisfactory at short wavelengths. It can be seen from equation 2 that as λ decreases E_λ increases, so that most of the radiant energy should be emitted at short wavelengths. Now both these methods correctly use the long accepted techniques of classical thermodynamics and mechanics; if the results contradict experimental observations, then the fundamental assumptions must be at fault. Max Planck's great achievement was the formulation of new assumptions.

He started work on this problem by deriving a purely empirical equation which would fit the experimental results. This equation was

$$E_\lambda \, d\lambda = \frac{c_1 \lambda^{-5}}{e^{(c_2/\lambda T)} - 1} \, d\lambda \qquad \dots (3)$$

Now it can be shown that equation 3 reduces to the Wien expression (equation 1) for small values of the product λT and to the Rayleigh-Jeans expression (equation 2) for large values of λT. For small values of λT the exponential function $e^{(c_2/\lambda T)}$ is very much larger than unity; the denominator in equation 3 thus reduces to $e^{(c_2/\lambda T)}$ and we get equation 1. Again, on expansion of the exponential function

$$e^{(c_2/\lambda T)} = 1 + \frac{c_2}{\lambda T} + \tfrac{1}{2} \left(\frac{c_2}{\lambda T} \right)^2 + \; \cdots$$

only the first two terms are significant for large values of λT and equation 3 reduces to equation 2. Planck now showed that any theory which was to produce equation 3 would have to relate the energy E with the frequency ν of the radiation by the equation $E = h\nu$, where h is a constant. (We shall in general refer to frequencies rather than wavelengths; they are related by $\nu = c/\lambda$ where c is the velocity of propagation.)

Such a relationship marks a complete departure from classical theory. The Rayleigh-Jeans method assumes that the electric oscillators associated with the electromagnetic radiation can have any energy values between zero and infinity. The Planck hypothesis states that the energy of these oscillators cannot vary continuously; they can only have definite amounts of energy, the so-called 'quanta', 0, $h\nu$, $2h\nu$, $\ldots$ $nh\nu$, where ν is the frequency, n an integer and h a universal constant, now known as Planck's constant. Any change in the energy of the oscillating system can only be in discrete amounts—one or more quanta. If this assumption is made, the constants c_1 and c_2 in the empirical equation 3 can be expressed in terms of fundamental constants, and the equation becomes

$$E_\lambda \, d\lambda = \frac{2\,\pi\,c^2\,h}{\lambda^5} \cdot \frac{1}{e^{(ch/\lambda kT)} - 1} d\lambda \qquad \ldots (4)$$

where c is the velocity of light and k the Boltzmann constant ($k = R/N$, where R is the gas constant and N the Avogadro number). By using the results of measurements of radiant energy and equation 4, Planck found $h = 6\cdot61 \times 10^{-27}$ erg sec. The modern value is $h = 6\cdot62554 \times 10^{-27}$ erg sec.

THE PHOTOELECTRIC EFFECT

Electrons may be emitted when light falls on a metal surface. In some cases (*e.g.* the alkali metals), light in the visible region of the spectrum can eject these so-called 'photoelectrons', but for most metals ultra-violet radiation must be used to produce the effect; there is a critical frequency, ν_0, for each metal, below which no photoelectrons are emitted. Experiments show that

(*a*) the energy of the photoelectrons is independent of the intensity, but proportional to the frequency of the incident radiation, and

(*b*) the number of photoelectrons emitted per second is proportional to the intensity of the incident radiation.

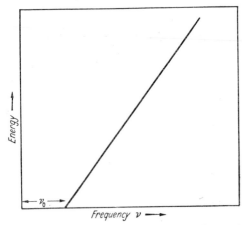

Figure 1.2. Variation of the maximum energy of photoelectrons with frequency of the incident radiation

Figure 1.2 shows how the maximum energy of the photoelectrons varies with the frequency of the incident radiation. The results may be expressed by the equation

$$\text{maximum energy} = \tfrac{1}{2}mv^2 = \textit{constant } (v - v_0) \quad \dots (5)$$

where m is the mass and v the velocity of the electron. Such a relationship cannot be derived from classical electromagnetic theory, however, for this predicts that the energy of the photoelectrons should vary with the intensity but should be independent of the frequency. Here, therefore, we have another example of the breakdown of classical radiation theory.

In 1905 Einstein showed that the difficulties could be resolved if the quantum postulates of Planck were applied to the photoelectric effect. Instead of regarding the incident light as radiation of frequency v, he considered it to be a stream of corpuscles, now called photons, each possessing energy hv, where h is the Planck constant. Each of these photons gives up its energy to an electron in the metal, and part of the energy is used in just removing the electron from the metal surface. The remainder appears as the kinetic energy of the photoelectron. Thus

$$hv = W + \tfrac{1}{2}mv^2 \quad \dots (6)$$

W is called the 'work function'; it represents the energy which is needed just to remove the electron from the metal surface. Comparing equation 6 with equation 5 representing the experimental results we can write

$$\tfrac{1}{2}mv^2 = hv - W = h(v - v_0) \quad \dots (7)$$

The value of h obtained from Millikan's (1917) experiments on the photoelectric effect, $6 \cdot 56 \times 10^{-27}$ erg sec, agrees well with that obtained from the radiation measurements.

THE BOHR THEORY OF THE HYDROGEN ATOM

The general acceptance of the quantum theory came with the development of a theory by Niels Bohr which accounted for some well known empirical relationships between the frequencies of the lines observed in the spectrum of hydrogen. When an electric discharge is passed through a tube containing hydrogen at low pressure, light is emitted, and its spectrum is found to consist of sharp lines. Four lines can be seen by the naked eye, and many others can be recorded photographically. *Figure 1.3* represents the lines as they appear on looking into the spectroscope eyepiece. (Note that the horizontal frequency scale is not linear; the dispersion of the instrument increases as the frequency increases.) It was shown by BALMER in 1885 that the wavelengths of many of these lines were given by a simple empirical formula

$$\lambda = \frac{bn^2}{n^2 - 4} \qquad \dots (8)$$

where b is a numerical constant and n an integer—*e.g.* 3, 4, 5, ... *etc.* This expression can be rearranged to give frequencies, ν, by using the relationship $\nu = c/\lambda$, where c is the velocity of light, giving

$$\nu = \frac{c(n^2 - 4)}{b \, n^2} = Rc\left\{\frac{1}{2^2} - \frac{1}{n^2}\right\}$$

where R, the so-called RYDBERG constant, is equal to $4/b$. Later work showed that the frequencies of all the lines in the hydrogen spectrum could be expressed by a general equation

$$\nu = Rc\left\{\frac{1}{n_1{}^2} - \frac{1}{n_2{}^2}\right\} \qquad \dots (9)$$

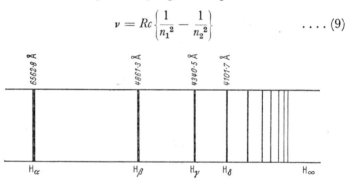

Figure 1.3. Atomic spectrum of hydrogen in the visible region

When $n_1 = 2$ and $n_2 = 3$, 4, 5, ... the Balmer lines are obtained; other series are obtained by giving n_1 other fixed values such as 1, or 3 or 4, and giving n_2 the values $(n_1 + 1)$, $(n_1 + 2)$, $(n_1 + 3)$ ... *etc.*

Now if we apply classical electromagnetic theory to the problem of the structure of the hydrogen atom, we come across a fundamental difficulty. The RUTHERFORD theory of atomic structure had postulated a small positively charged nucleus containing most of the mass of the atom, surrounded by electrons. A simple mechanical model of this system is that of electrons moving in orbits around the nucleus, like planets round the sun in the solar system. The negatively charged electrons are held in their orbits by electrostatic attraction to the positively charged nucleus. This attraction varies inversely as the square of the distance, just as the gravitational force varies in the case of the solar system ; the mathematical theory developed by KEPLER and NEWTON for the motion of heavenly bodies should therefore apply to the motion of electrons in atoms.

There is this difficulty, however. Classical electromagnetic theory states that an accelerating electric charge must radiate energy, so that an electron moving in an orbit around a nucleus must therefore radiate energy, and move in a decreasing spiral path until it finally disappears into the nucleus. The quantum theory of radiation avoids this difficulty.

Planck's quantum theory introduced two important ideas:

(a) that electric oscillators can only possess discrete quanta of energy, and

(b) that radiation is emitted only when the oscillator changes from one quantized state to another of lower energy.

Bohr now (1913) adapted these ideas to the quite different system of the hydrogen atom, and postulated that the motion of the electron was restricted to a number of discrete circular orbits, with the nucleus at the centre. An electron moving in one of these orbits has a constant energy, and radiation is emitted only when the electron moves into an orbit of lower energy. Bohr also adapted Planck's expression for the relationship between frequency and energy, for he assumed that the frequency of the emitted radiation is given by

$$\nu = (E_2 - E_1)/h$$

where E_2 and E_1 are the energies of the electron in two different orbits, and h is Planck's constant. The size of the orbits was determined by an arbitrary assumption that the angular momentum of the electron about the nucleus is an integral multiple of $h/2\pi$. A quantizing condition of this form had previously been put forward by J. W. NICHOLSON (1912), and Bohr was able to use it, together with the postulates already mentioned, to derive the empirical Balmer Law in which the value of the Rydberg constant determined by theory was in satisfactory agreement with the experimental value (see Chapter 2).

The success of the quantum theory in removing difficulties encountered by classical theory in three such different branches of physics led to its rapid acceptance as the most satisfactory method of discussing phenomena on the atomic scale. The variation of the specific heat of solids with temperature is another important case in which the quantum theory has been applied with success. There is, therefore, ample experimental justification for taking the quantum theory as the foundation of a modern discussion of chemical behaviour, in spite of the drastic break such a treatment makes with traditional methods in this field.

ENERGY UNITS AND THE MAGNITUDES OF ENERGY QUANTA[4]

In later chapters we shall often need to refer to the energies of atomic systems. We therefore include here a short section on the units in which energy is often expressed, and the method of converting one unit to another.

In physics, energy is usually expressed in ergs or in joules:

$$1 \text{ joule} = 10^7 \text{ erg}$$

Chemists, however, are accustomed to discussing the energy changes in chemical reactions in calories or kilocalories, using the definition

$$1 \text{ calorie} = 4 \cdot 1840 \text{ joules}$$

Very large energy values (*e.g.* in nuclear reactions) are often quoted in electron volts (eV), where the electron volt is defined as the energy acquired by an electron when its potential changes by 1 volt. Using the values 1 e.s.u. of potential $= 299 \cdot 796$ volts and electronic charge $e = 4 \cdot 80296 \times 10^{-10}$ e.s.u., and the fact that the energy acquired by a charge e (e.s.u.) when the potential changes by v e.s.u. is ve erg, we obtain

$$1 \text{ electron volt} = \frac{4 \cdot 80296 \times 10^{-10}}{299 \cdot 796} = 1 \cdot 6021 \times 10^{-12} \text{ erg}$$

Energy may also be expressed in frequency units if we make use of the quantum relationship $E = h\nu$, where h is the Planck constant and ν the frequency (vibrations per second). Spectroscopic frequencies, however, are usually expressed in 'wave-number' units, ν', where

$$\nu' = \frac{1}{\lambda} \left(= \frac{\nu}{c} \right)$$

measured in reciprocal centimetres (cm^{-1}). Substitution of $h = 6 \cdot 62554 \times 10^{-27}$ and $c = 2 \cdot 99796 \times 10^{10}$ in $E = hc\nu'$ gives

$$1 \text{ cm}^{-1} = 1 \cdot 98629 \times 10^{-16} \text{ erg}$$

This conversion factor relates to the emission or absorption of a quantum of energy by a single atom, and the corresponding factor for a gramme-molecule is obtained by multiplying by the Avogadro number, N $(= 6 \cdot 0226 \times 10^{23})$. The spectroscopic, thermal and electrical units of energy are then related as follows:

$$1 \text{ cm}^{-1} = 2 \cdot 8591 \text{ cal mol}^{-1} = 1 \cdot 2398 \times 10^{-4} \text{ eV atom}^{-1}$$
$$1 \text{ eV} = 23 \cdot 0612 \text{ kcal mol}^{-1}$$

As an example of these interconversions we conclude this section with the numerical values of the energies corresponding to radiation of three different wavelengths, one in the infra-red, one in the visible, and one in the ultra-violet region of the spectrum. The wavelengths are measured in Ångström units, where $1\text{Å} = 10^{-8}$ cm.

Wavelength Å	Wave-number cm^{-1}	Energy cal mol^{-1}	Energy eV
10,000	10,000	28,591	1·2398
5000	20,000	57,182	2·4796
2000	50,000	142,955	6·1990

REFERENCES

[1] PLANCK, M. *Ann. Phys., Lpz.* 1 (1900) 69
[2] EINSTEIN, A. *Ann. Phys., Lpz.* 17 (1905) 132
[3] BOHR, N. *Phil. Mag.* 26 (1913) 476, 857
[4] COHEN, E. R. and DUMOND, J. W. M. *Phys. Rev. Ltrs.* 1 (1958) 291, 382; *see also Tech. News Bull. natn. Bur. Stand.* 47 (1963) 175

For further reading—

RICHTMEYER and KENNARD. *Introduction to Modern Physics*, McGraw-Hill, New York, 1947

2

ATOMIC STRUCTURE AND THE
QUANTUM THEORY—I

WE saw in the last chapter that empirical relationships between the frequencies of lines in the hydrogen atom spectrum can be explained by the use of the quantum theory. We shall now discuss the Bohr theory of the hydrogen atom in some detail, and outline the ways in which the theory may be modified to include elliptical as well as circular orbits, and atoms other than hydrogen. The great advantage of the Bohr theory lies in the fact that it uses a simple model of the atom that can be readily visualized by the non-mathematician, and that can, moreover, explain the principal features of the spectra of many atoms.

THE BOHR THEORY—CIRCULAR ORBITS

We consider an electron of mass m and charge e moving in a circular orbit of radius r with velocity v about a nucleus of charge Ze (Z is the atomic number of the nucleus, *i.e.* the number of protons in it, and e is the magnitude of the proton charge). We shall assume in the first instance that the mass of the nucleus is so great compared with the mass of the electron that we can consider the nucleus to be stationary. The total energy E of the electron in this orbit is the sum of its kinetic energy, $\frac{1}{2}mv^2$, and its potential energy. The potential energy can only be defined by reference to some arbitrary energy zero; it is convenient in this case to take the energy as zero when the electron is at rest at infinite distance from the nucleus. The potential energy of the electron at a distance r from the nucleus is then $- Ze^2/r$.

The negative sign for this potential energy sometimes causes difficulty; it arises from the choice of energy zero. It may help to consider a simple analogy, the measurement of the height of a mountain or the depth of an ocean. If we define sea level to be our arbitrary zero of height, then mountains have a positive height and ocean beds a negative height. We could measure all heights from the lowest level, the sea bed, thus giving them all positive values; it would be very difficult, however, to find a reference level applicable to all situations. In the case of atoms, the choice of our energy

12

zero gives a reference level applicable to all atoms, and this more than compensates for the inconvenience of thinking in terms of negative energies. The important thing is to be quite clear that if, for example, an atom had two possible energy states, say — 20,000 and — 25,000 cal, the state of 'lower' energy is the latter, although the numerical magnitude is greater. The total energy of the electron is therefore given by

$$E = \tfrac{1}{2}mv^2 - Ze^2/r \qquad \dots (10)$$

Now the centrifugal force, mv^2/r, tending to drive the electron away from the nucleus must be balanced by the electrostatic force of attraction, Ze^2/r^2, between the electron and the positively charged nucleus. Thus

$$mv^2/r = Ze^2/r^2 \qquad \dots (11)$$

and equation 10 becomes

$$E = Ze^2/2r - Ze^2/r = - Ze^2/2r = - mv^2/2 \qquad \dots (12)$$

We now introduce the first of the Bohr assumptions—*viz.* that the electron can only revolve in an orbit of radius r if its angular momentum about the nucleus, mvr, is equal to $nh/2\pi$, where n is an integer and h the Planck constant. Substituting $mvr = nh/2\pi$ in equation 11, we get

$$v = 2\pi e^2 Z/nh \qquad \dots (13)$$

and substituting this value for v in equation 12

$$E = - 2\pi^2 me^4 Z^2/n^2 h^2 \qquad \dots (14)$$

Thus the stable state of the hydrogen atom, *i.e.* the state of *lowest* energy, is that for which n is equal to unity, while states of higher energy are characterized by larger values of n; n is called the 'principal quantum number'. The radii of the allowed orbits can be obtained by substituting the value of E given by equation 14 in equation 12

$$r = - Ze^2/2E = n^2 h^2/4\pi^2 me^2 Z \qquad \dots (15)$$

From this we see that the state of lowest energy, for which $n = 1$, is also the state with the orbit of smallest radius. Substitution of the known values for h, m, and e in equation 15 gives the radius of the orbit of the stable state of the hydrogen atom to be $0 \cdot 53$ Å, in reasonable agreement with the radius obtained from kinetic theory methods.

We can now apply the second Bohr assumption to determine the frequencies in the atomic spectrum. If an electron passes from an orbit of principal quantum number n_2 and energy E_2 to an orbit of lower energy E_1 and smaller principal quantum number n_1, the energy radiated, ΔE, will have an associated frequency ν given by $\Delta E = h\nu$. Thus, using equation 14, we have

$$\nu = \frac{\Delta E}{h} = \frac{E_2 - E_1}{h} = \frac{2\pi^2 m e^4 Z^2}{h^3}\left\{\frac{1}{n_1^2} - \frac{1}{n_2^2}\right\} \quad \dots(16)$$

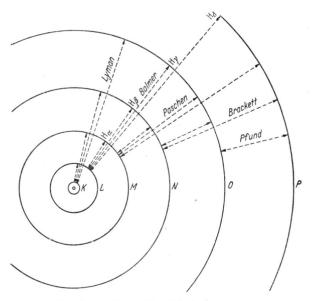

Figure 2.1. Bohr orbits of the hydrogen atom

If we express the frequency in wave numbers, ν', then

$$\nu' = \frac{\nu}{c} = \frac{2\pi^2 m e^4 Z^2}{c h^3}\left\{\frac{1}{n_1^2} - \frac{1}{n_2^2}\right\} = RZ^2\left\{\frac{1}{n_1^2} - \frac{1}{n_2^2}\right\} \quad \dots(17)$$

where R, the Rydberg constant, is written for $2\pi^2 me^4/c\,h^3$.

The fact that the theory produces an equation of the same form as the earlier empirical one (equation 9) does not in itself show that the theory is correct; indeed, the form of equation 17 follows inevitably from the nature of the Bohr assumptions. What is important is that the value of R obtained experimentally, using the empirical equation, agrees very closely with that calculated by inserting appropriate values for the quantities in equation 17. Thus the experimental value of R for hydrogen is 109,677·58 cm^{-1} and the calculated value 109,737 cm^{-1}.

We can see from equation 17 that different series of spectrum lines will be obtained by suitably varying the values of n_1 and n_2. Thus with $n_1 = 1$ and $n_2 = 2, 3, 4$ . . . we get lines in the ultra-violet region of the spectrum known as the Lyman series; $n_1 = 2$ and $n_2 = 3, 4, 5$ . . . gives us the Balmer series in the visible region; $n_1 = 3$ and $n_2 = 4, 5, 6$ . . . gives the Paschen series, and $n_1 = 4$, $n_2 = 5, 6, 7$ . . . the Brackett series of lines, both in the infra-red. The transitions between the various orbits are shown diagrammatically in *Figure 2.1*.

EXTENSIONS OF THE BOHR THEORY

Two refinements to the simple theory can now be made. If the motion of the nucleus is taken into account, equation 17 becomes

$$\nu' = \frac{2\pi^2 e^4 \mu Z^2}{ch^3} \left\{ \frac{1}{n_1^2} - \frac{1}{n_2^2} \right\}$$

where m, the electron mass, has been replaced by the so-called 'reduced mass', μ; this is defined by $\mu = mM/(m + M)$, where M is the mass of the nucleus. Thus for hydrogen $(Z = 1)$

$$\nu' = \frac{2\pi^2 e^4}{ch^3} \frac{m M_H}{m + M_H} \left\{ \frac{1}{n_1^2} - \frac{1}{n_2^2} \right\}$$

$$= R\infty \frac{M_H}{m + M_H} \left\{ \frac{1}{n_1^2} - \frac{1}{n_2^2} \right\}$$

where $R\infty = 2\pi^2 e^4 m/ch^3$. The value of R obtained by experiment therefore varies slightly with the nuclear mass. It is for this reason that the existence of some isotopes can be demonstrated from atomic spectrum observations. In 1931, UREY, BRICKWEDDE and MURPHY discovered deuterium by observing the Balmer lines given by the gas obtained from the residue left after the evaporation of a large volume of liquid hydrogen. A faint satellite of the hydrogen H_α line was detected at a separation of 1·79 Å, corresponding to the presence of a nucleus of mass double that of the hydrogen nucleus. The change in R is not very large; values for hydrogen, deuterium, and a nucleus of infinite mass are as follows: $R_H = 109,677·58$ cm^{-1}, $R_D = 109,707·42$ cm^{-1}, $R\infty = 109,737·30$ cm^{-1}.

The second refinement, the extension of the theory to the case of elliptical orbits, was made by SOMMERFELD. The quantum relationship $mvr = nh/2\pi$ defines the radii of possible circular orbits in terms of a single quantum number n. A more complex quantum restriction is required to define an elliptical orbit; two quantum numbers, n and k, are involved, where n can take any integral value as before, and k is allowed to take values from 1 to n. Sommerfeld showed that if account was taken of the change in electron mass with velocity, as required by the relativity theory, electrons in orbits of the same value of n, but with different k values, had slightly different energies. The ratio n/k is equal to the ratio of the semi-major to the semi-minor axis of the ellipse. If $n = 4$, $k = 4, 3, 2, 1$; $k = 4$ would correspond to a circular orbit, and $k = 1$ to a very narrow, elongated elliptical orbit. The value $k = 0$ was excluded on the grounds that it would correspond to a linear motion of the electron, passing through the nucleus. Later work showed that this introduction of the k quantum number was not completely satisfactory, and we shall see (in Chapter 5) that the new quantum theory replaces k by l, where l can take the values $0, 1, 2, \ldots (n - 1)$. It will avoid confusion if we make the change at this stage; we shall therefore always refer in future to the l quantum number, where $l = k - 1$.

Thus, if E_3 represents the energy of an electron in a circular orbit defined by the principal quantum number 3, $E_{3, 0}$, $E_{3, 1}$, $E_{3, 2}$ represent the energies of the electron in elliptical or circular orbits with the same principal quantum number. In the case of hydrogen, these energy differences are very small for orbits of the same n value. The additional orbits will, however, be responsible for the appearance of additional lines in the atomic spectrum, since radiation appears as the result of electron transitions between states of different energy. Observation of the hydrogen spectrum with an instrument of high dispersing power reveals that lines which appear single under low dispersion do, in fact, consist of two or more lines very close together.

The theory can be extended to deal with atoms other than hydrogen. In these cases, the energies of electrons in orbits of the same n value, but different l values, may be very different. *Figure 2.2* illustrates the way in which the different energy states may be represented diagrammatically. Energy is plotted on the vertical scale, with the zero, corresponding to complete removal of the electron from the atom, at the top. The energy of a particular orbit is shown by a horizontal line, and it is customary to speak of the 'energy levels' of an electron in an atom. Levels corresponding to orbits of different l values are arranged in different columns, and we thus get a series of levels for which $l = 0, 1, 2, 3, \ldots$ *etc.*; spectroscopists refer to these as S, P, D, and F* levels, respectively. Electron transitions between different levels, resulting in the production of spectrum lines, can be shown on energy-level diagrams by lines marked with arrows. We can conveniently refer to a level for which $n = 2$ and $l = 1$ as a 2P level, and one for which $n = 3$

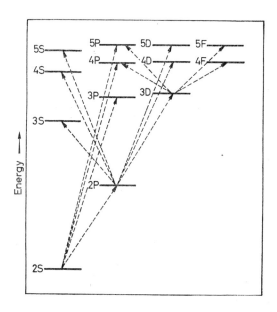

Figure 2.2. Energy level diagram for lithium

and $l = 2$ as a 3D level; a transition between these two levels as shown in *Figure 2.2* would be written as 2P → 3D. Now it is observed that the number of lines appearing in the spectrum of an element is much smaller than would be expected from the number of energy levels. An empirical limiting factor known as a 'selection rule' is therefore introduced to restrict the possible transitions to those for which l changes by ± 1.

Thus a 2P → 3P transition, or a 2S → 3D transition, would be forbidden by the selection rule, since in one case l remains unchanged, while in the other case, l changes from 0 to 2.

* These are selected as the initial letters of the words sharp, principal, diffuse and fundamental, thus taking over a terminology used by spectroscopists in describing series of lines observed in atomic spectra.

THE SPINNING ELECTRON

The principal features of the spectra of many atoms can thus be explained by the introduction of two quantum assumptions, involving two quantum numbers, and a selection rule. New spectrum lines appear when the atoms are placed in a strong external magnetic field. This is called the ZEEMAN effect. The plane containing the electron orbit can only take up certain specific orientations with respect to the direction defined by the external field. Each of these orientations is associated with a third quantum number, m, the 'magnetic quantum number', which can take the values $-l$, $-(l-1)$, $\ldots -2, -1, 0, 1, 2 \ldots (l-1), l$.

The existence of the 'double' lines in the spectra of the alkali metals was attributed by GOUDSMIT and UHLENBECK in 1925 to the axial *spin* of the electron. A simplified account of the way in which this property leads to new energy levels can be given if we

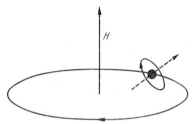

Figure 2.3. Orbital and spin motion of an electron

remember that a spinning electron behaves as a small magnet. Now the electron as it moves around its orbit will produce a magnetic field just as an electric current in a coil of wire produces a field. The arrow marked H in *Figure 2.3* represents this field. Since the electron because of its axial spin behaves as a small magnet, there will be an interaction between the two magnetic fields. The field produced by the axial spin will either reinforce the field H or oppose it, according as the direction of the spin is clockwise or anti-clockwise. The interaction will produce energy changes; a single energy level, representing a non-spinning electron moving in an orbit, becomes two energy levels very close together. Additional electron transitions are therefore possible, and new lines appear in the spectrum. Goudsmit and Uhlenbeck showed that the spectroscopic observations required that the angular momentum associated with the spin of the electron be given by $m_s \cdot h/2\pi$, where m_s is called the 'spin quantum number' and can have the value $+\frac{1}{2}$ or $-\frac{1}{2}$.

We have now moved a long way from the Bohr theory: each observed increase in complexity of the spectrum requires a new arbitrary assumption. This is an unsatisfactory situation, and a new theoretical approach is evidently needed. Nevertheless the principal results of the simple theory, the existence of energy levels in the atoms, and the assignment of four quantum numbers to each electron, remain of fundamental importance.

REFERENCE

[1] UREY, H. C., BRICKWEDDE, F. G., and MURPHY, G. M. *Phys. Rev.* 40 (1932) 1

For further reading—

HERZBERG, G. *Atomic Spectra and Atomic Structure*, New York, Dover Publications, 1944

ATOMIC STRUCTURE AND THE
QUANTUM THEORY—II

PARTICLES AND WAVES

WE saw in Chapter I that, although the phenomena of the inter-
ference and diffraction of light required the use of a wave theory of
radiation, the photoelectric effect could only be explained by
assuming that light travelled as particles (photons). Now many of
the phenomena of optics—reflection and refraction for example—
can be quite adequately explained by a particle theory; it is only
when experiments are done with apertures or obstacles of very
small dimensions that diffraction effects are obtained, and a wave
theory is needed. We thus get a duality of behaviour; light radiation
can be considered either as a stream of photons or as a wave motion.
It was pointed out in 1924 by DE BROGLIE[1] that a similar duality
might exist in the case of the electron. All the experiments per-
formed on electrons from their discovery up to this time could be
explained by assuming they were small particles of a certain mass
and charge, but de Broglie showed, by arguments based on the
theory of relativity, that if an electron of momentum p could be
described by a wave theory in terms of a wavelength λ, then:

$$\lambda = h/p \qquad \dots \text{ (18)}$$

where h is the Planck constant. We can use equation 18 to obtain
a numerical value for λ. For an electron of velocity 6×10^7
cm sec^{-1}, corresponding to an energy of about 1 eV, λ is of the order
of 10 Å. Electrons of greater energy, and consequently greater
momentum, will have a smaller associated wavelength. These
wavelengths are of the same order of magnitude as the spacings of
atoms in crystals, and it was realized that it might be possible to use
crystals as diffraction gratings for electrons, just as they can be used
to diffract X-rays whose wavelengths are also of this order. Such
diffraction effects were in fact observed in 1927. DAVISSON and
GERMER[2] observed the reflection of an electron beam from the
surface of a metal, and THOMSON[3] and REID photographed the
diffraction rings produced when the electrons passed through a very

thin metal foil. In each case the experimental results fitted equation 18. With this experimental evidence for the behaviour of electrons as waves it was possible to attempt to describe the motion of electrons in atoms in terms of a wave theory; indeed the 'new' quantum theory is the application of the techniques of wave mechanics to atomic systems.

We must now consider what is required from a new quantum theory of the atom. The Bohr theory succeeded because it provided both a simple physical picture of the behaviour of electrons in atoms, and a means of calculating electron energies. Any new theory must do all this, but it must also account for the diffraction of electrons, and must reduce the number of arbitrary assumptions that the older theory had to make. We shall first of all discuss the problem of using a wave theory to describe the position of particles. The photon theory of light relates photon density with light intensity; a large number of photons per unit volume produces an intense spot of light. In wave theory, intensity is measured by the square of a quantity called the 'amplitude', and this quantity can be obtained from the solution of a 'wave equation' which describes the particular system. If, therefore, we use a wave theory to calculate the variations in intensity in a diffraction experiment, we can express the results in terms of a particle theory by equating the square of the amplitude with the particle density. The new quantum theory describes the behaviour of a beam of electrons in the same way. The dark rings on the photographic plate in an electron diffraction experiment reveal positions of high electron density; these densities are related to the square of the amplitude factor obtained by solving a suitable wave equation. It is not quite so easy to describe the behaviour of a single electron since the concept of particle density seems to lose most of its meaning when there is only one electron available.

THE UNCERTAINTY PRINCIPLE

There is, moreover, another fundamental difficulty, which lies at the root of all atomic theory. This difficulty, usually referred to as the 'HEISENBERG uncertainty principle', is that it is impossible to state, simultaneously and precisely, both the position and the momentum of an electron. Suppose we devise a hypothetical experiment to measure the position and velocity of an electron. We could set up two 'γ-ray' microscopes that could 'see' electrons, and measure the time taken for the electron to pass from one to the other. P, in *Figure 3.1,* represents the electron. It can only be observed if a photon incident upon it is scattered into the aperture of the microscope, *i.e.* within the cone of angle 2α. Now a photon of frequency ν will have an associated wavelength $\lambda = c/\nu$, and by equation 18, $c/\nu = h/p$, whence $p = h\nu/c$.

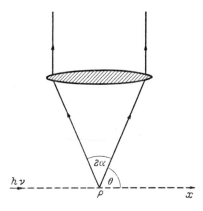

Figure 3.1. The 'γ-ray' microscope

If the photon is scattered in a direction making an angle θ with the x axis, the electron will receive a component of momentum along the x axis of

$$\frac{h\nu}{c} (1 - \cos \theta)$$

and since the electron will be detected for any value of the angle θ between $90° \pm \alpha$, the momentum may have any value between

$$\frac{h\nu}{c} [1 - \cos (90 - \alpha)]$$

and

$$\frac{h\nu}{c} [1 - \cos (90 + \alpha)]$$

i.e. between

$$\frac{h\nu}{c} (1 - \sin \alpha)$$

and

$$\frac{h\nu}{c} (1 + \sin \alpha)$$

If we define this spread of values as the 'uncertainty' in the value of p, and denote it by Δp, then

$$\Delta p = \frac{2h\nu}{c} \sin \alpha$$

22

We could try to reduce this uncertainty by making α small, *i.e.* by using a microscope of smaller aperture, but the accuracy with which an object can be located by a microscope is defined by the Rayleigh equation for the resolving power,

$$\Delta x = \frac{c}{\nu \sin \alpha}$$

where Δx is the uncertainty in x, the co-ordinate defining the position of the electron. Thus a smaller aperture, while decreasing the uncertainty in the momentum, would increase the uncertainty in position. In this experiment

$$\Delta x \ . \ \Delta p = \frac{2h \nu}{c} \sin \alpha \ . \ \frac{c}{\nu \sin \alpha} = 2h \qquad \ldots .(19)$$

In general, the product $\Delta x \ . \ \Delta p$ is of the order of magnitude of the Planck constant, h. This is one way of expressing the Heisenberg uncertainty principle (1927).

We can illustrate its importance by a rough calculation. Suppose that we could locate the position of an electron with an 'uncertainty' of $0 \cdot 001$ Å, *i.e.* $\Delta x = 10^{-11}$ cm. Substituting values for Δx and h in equation 19 then gives

$$\Delta p \sim 2 \times 6 \cdot 6 \times 10^{-27}/10^{-11} = 13 \cdot 2 \times 10^{-16} \ \text{g. cm. sec}^{-1}$$

This uncertainty in momentum would be quite negligible in macroscopic systems, but it is far from negligible in systems containing electrons, since we are then dealing with masses of the order of 10^{-27} g. Precise statements of the position and momentum of electrons have to be replaced by statements of the probability that the electron has a given position and momentum. The introduction of probabilities into the description of electronic behaviour is a direct consequence of the operation of the uncertainty principle; a small uncertainty in position implies a high probability that the electron is at a given point. This probability concept can be further illustrated if we reconsider the electron diffraction experiments of Thomson (page 20), in which the diffraction rings obtained corresponded to regions of high electron density. If a single electron is sent through the diffraction apparatus it obviously cannot interfere with itself to give a diffraction pattern, and the Heisenberg principle tells us that we cannot follow its course precisely. We can say, however, that there is a certain probability that it will take a particular path, and that the electron is most likely to be found in those regions where we get the greatest electron density in experiments using beams of electrons. Thus a high intensity in a diffraction experiment, measured by the square of an amplitude factor in a wave equation, can be related to a high probability that an electron is in a unit volume around a given point.

We shall make use of both of these concepts—electron density and electron probability.

WAVES AND WAVE EQUATIONS

It is easy to visualize a wave motion in terms of the disturbance that spreads over the surface of a pond when a stone is dropped into

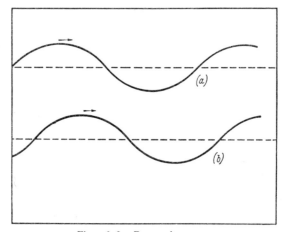

Figure 3.2. Progressive waves

it, or in terms of the vibrations of a violin string; it is not so easy to get a physical picture of the sound waves in an organ pipe, or the electro-magnetic waves of light radiation. All these wave motions, however, are characterized by a transmission of energy from one point to another without permanent displacement of the intervening medium; thus waves can be made to travel down a stretched string, but the string itself does not undergo permanent translation. These wave motions can, moreover, all be described by wave equations of similar form. Let us consider a simple case, the wave motion produced by moving up and down one end of a long horizontal string. The appearance (or 'profile') of the string at some instant will be as shown in *Figure 3.2(a)*, and at a later instant by

WAVES AND WAVE EQUATIONS

Figure 3.2(b). The disturbance is thus transmitted along the string in the form of a progressive wave. The equation for this wave motion, which can be shown to be

$$\frac{\partial^2 \phi}{\partial x^2} = \frac{1}{c^2} \cdot \frac{\partial^2 \phi}{\partial t^2} \qquad \dots (20)$$

has a solution of the form

$$\phi = a \sin 2\pi(x/\lambda - \nu t) \qquad \dots (21)$$

where ϕ is the amplitude of the disturbance at a distance x along the string, λ is the wavelength, ν the frequency, a a constant (the maximum value of the amplitude), and c the velocity of the progression. (The reader can check that equation 21 is a solution of equation 20 by appropriate differentiation of equation 21 and substitution in equation 20.) Equation 20 is 'linear'—*i.e.* the function ϕ and its differential coefficients are always of the first order. An important characteristic of such linear equations is that if ϕ_1 and ϕ_2 are any two solutions, the linear combination $a_1\phi_1 + a_2\phi_2$, where a_1 and a_2 are arbitrary constants, is also a solution. This can be seen by substitution in equation 20; thus

$$\frac{\partial^2 \phi_1}{\partial x^2} = \frac{1}{c^2} \cdot \frac{\partial^2 \phi_1}{\partial t^2}, \quad \frac{\partial^2 \phi_2}{\partial x^2} = \frac{1}{c^2} \cdot \frac{\partial^2 \phi_2}{\partial t^2}$$

and

$$\frac{\partial^2(a_1\phi_1 + a_2\phi_2)}{\partial x^2} = a_1 \frac{\partial^2 \phi_1}{\partial x^2} + a_2 \frac{\partial^2 \phi_2}{\partial x^2}$$

$$= \frac{1}{c^2}\left\{ a_1 \frac{\partial^2 \phi_1}{\partial t^2} + a_2 \frac{\partial^2 \phi_2}{\partial t^2} \right\}$$

$$= \frac{1}{c^2} \cdot \frac{\partial^2(a_1\phi_1 + a_2\phi_2)}{\partial t^2}.$$

This is an example of the important 'Principle of Superposition' which is used extensively in the quantum theory of valency (*see* page 73). We can apply the principle at this stage to the problem of the vibration of a string stretched between two fixed points. A progressive wave can be made to travel down the string from left to right, the amplitude being given by the equation

$$\phi_1 = a \sin 2\pi(x/\lambda - \nu t) \qquad \dots (22)$$

When the wave reaches the fixed end of the string it is reflected, and travels back from right to left, with amplitude ϕ_2 given by

$$\phi_2 = a \sin 2\pi(x/\lambda + \nu t) \qquad \dots (23)$$

This wave will interfere with the wave travelling from left to right, and the resulting wave motion has an amplitude ϕ given by

$$\phi = \phi_1 + \phi_2 = a \sin 2\pi(x/\lambda - vt) + a \sin 2\pi(x/\lambda + vt)$$

which by simple trigonometry gives

$$\phi = 2a \sin 2\pi x/\lambda \ . \ \cos 2\pi vt \qquad \ldots(24)$$

Inspection of equation 24 shows that ϕ will be zero when $\sin \dfrac{2\pi x}{\lambda}$ is zero, i.e. when

$$\frac{2\pi x}{\lambda} = n\pi \text{ and } x = \frac{n\lambda}{2} \qquad \ldots(25)$$

where n is an integer. These waves are called 'standing' or 'stationary' waves; the amplitude is always zero at the particular values of x given by equation 25, whereas with 'progressive' waves the amplitude at any value of x is continually changing. We shall find that wave equations appropriate to the description of the behaviour of electrons in atoms are analogous to those describing 'stationary' waves in ordinary mechanics, and it will therefore be useful to consider this example in rather more detail.

We can for convenience write equation 24 in the form

$$\phi = f(x) \ . \ \cos 2\pi vt$$

Simple differentiation then gives

$$\frac{\partial \phi}{\partial x} = \cos 2\pi vt \ . \ \frac{\partial f(x)}{\partial x}$$

$$\frac{\partial^2 \phi}{\partial x^2} = \cos 2\pi vt \ . \ \frac{\partial^2 f(x)}{\partial x^2}$$

$$\frac{\partial \phi}{\partial t} = -f(x) \ . \ 2\pi v \sin 2\pi vt$$

$$\frac{\partial^2 \phi}{\partial t^2} = -f(x) \ . \ 4\pi^2 v^2 \cos 2\pi vt$$

Substitution of these values in equation 20 gives

$$\cos 2\pi vt \ . \ \frac{\partial^2 f(x)}{\partial x^2} = \frac{1}{c^2} \ . \ (-f(x) \ . \ 4\pi^2 v^2 \cos 2\pi vt)$$

i.e.

$$\frac{d^2 f(x)}{dx^2} = -\frac{4\pi^2 v^2}{c^2} f(x) \qquad \ldots(26)$$

if c is constant. We have thus eliminated the variable t from our original wave equation 20, and equation 26 does not, therefore, contain *partial* differentials.

Now $c = \lambda\nu$, where λ is the wavelength and ν the frequency, and we can therefore write equation 26 in the form

$$\frac{\mathrm{d}^2 f(x)}{\mathrm{d}x^2} = -\frac{4\pi^2}{\lambda^2} f(x) \qquad \dots(27)$$

It is important to realize that not every possible solution of equation 27 is a physically acceptable one. We can perhaps best make this clear by considering the solution of the simplest of all differential equations

$$\mathrm{d}y/\mathrm{d}x = 1$$

The solutions of this are $y = x + a$, where a, the arbitrary integration constant, can have any value we choose to give it. A graphical representation of these solutions would be an infinite number of parallel straight lines; a is given by the intercept on the y axis in

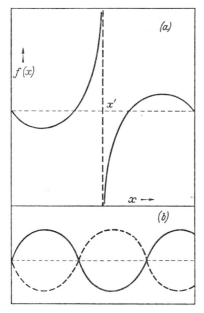

Figure 3.3. Unacceptable solutions of the wave equation for vibrations of a stretched string

27

each case. In the same way the solution of equation 27 involves an arbitrary constant, and we get a large number of solutions $f(x_1)$, $f(x_2)$, $f(x_3)$, *etc.* The only acceptable ones will be those that satisfy what are known as 'continuity' and 'boundary' conditions which state that $f(x)$ must be continuous, finite, and single-valued, for every value of x between given boundaries.

Thus the curve of *Figure 3.3(a)* represents a function which would not be an acceptable solution of the wave equation, since there is a discontinuity at $x = x'$, and since, in addition, the function has an

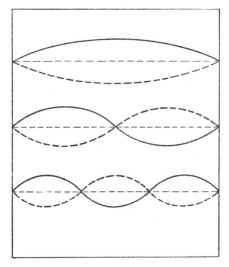

Figure 3.4. Acceptable solutions of the wave equation for vibrations of a stretched string

infinite value at this point. The 'single-valued' limitation is self-explanatory, for the amplitude can have only one value at a given position x'. The boundary conditions are imposed by the physical constraints acting on the vibrating system; thus, in the case of the stretched string, $f(x)$ must be zero at each end. *Figure 3.3(b)* represents a solution that is not acceptable since $f(x)$ is not zero at the right-hand fixed end, but some acceptable solutions are shown in *Figure 3.4.* Functions which are acceptable solutions in this sense are called 'Eigenfunctions'; the corresponding values of λ are called 'Eigenvalues'. (These words are German–English hybrids; the German expression 'eigen-funktion', or the English translation 'characteristic function' would be preferable, but the hybrids are now too firmly established.) We shall often, however, refer simply to 'wave-functions', implying that these are, in fact, *eigen*-functions.

THE WAVE EQUATION FOR ELECTRONS

We have now reached the stage where we can formulate a wave equation for electrons. Historically, the wave mechanical theory of the hydrogen atom began in 1927, when SCHRÖDINGER[4] wrote down a wave equation to describe electron behaviour, the form of the equation being determined by mathematical intuition. This equation was then solved to give numerical values for quantities that could be determined experimentally, and agreement between calculation and experiment was then taken to confirm the correctness of the postulated wave equation. Experience shows, however, that chemists usually respond without enthusiasm to the bald initial statement—'. . . the behaviour of electrons in atoms can be represented by the differential equation. . . '. We shall therefore develop a simple treatment which is certainly not rigorous in the mathematical sense of the word, but which may help to show that the form of the postulated wave equation is not entirely arbitrary. First of all we must realize that the electron in, say, a hydrogen atom is under certain constraints imposed by the attraction of the nucleus. The wave equation that describes the electron's motion must therefore be an equation analogous to that used to describe a standing wave system, since this also represents a constrained system, *e.g.* the vibrating string fixed at its ends. If the electron moved in one dimension only, the appropriate equation would be that of equation 27 which we now write

$$\frac{d^2\psi}{dx^2} = -\frac{4\pi^2}{\lambda^2}\psi \qquad \ldots(28)$$

where ψ has been written for $f(x)$. This equation can be extended to describe motion in three dimensions, when it becomes:

$$\frac{\partial^2\psi}{\partial x^2} + \frac{\partial^2\psi}{\partial y^2} + \frac{\partial^2\psi}{\partial z^2} + \frac{4\pi^2}{\lambda^2}\psi = 0 \qquad \ldots(29)$$

in which ψ is a function of the Cartesian coordinates x, y, and z. This may be written more concisely as

$$\nabla^2\psi + \frac{4\pi^2}{\lambda^2}\psi = 0 \qquad \ldots(30)$$

where $\nabla^2\psi$ is written for

$$\frac{\partial^2\psi}{\partial x^2} + \frac{\partial^2\psi}{\partial y^2} + \frac{\partial^2\psi}{\partial z^2}$$

3

We now use the de Broglie relationship, equation 18 ; if the electron has an associated wavelength λ, given by

$$\lambda = \frac{h}{p} = \frac{h}{mv}$$

then substitution in equation 30 gives

$$\nabla^2\psi + \frac{4\pi^2\, m^2\, v^2}{h^2}\, \psi = 0 \qquad \ldots . (31)$$

We wish to use our wave equation to calculate values for the energy states of the hydrogen atom, so we make use of the fact that kinetic energy, $\frac{1}{2}mv^2$, is given by $E - V$, where E is the total, and V the potential energy. Substituting

$$v^2 = \frac{2}{m}\,(E - V)$$

in equation 31, we get

$$\nabla^2\psi + \frac{8\,\pi^2\, m}{h^2}\,(E - V)\,\psi = 0 \qquad \ldots . (32)$$

which is the celebrated Schrödinger equation describing the behaviour of the electron in the hydrogen atom. We emphasize again that the above treatment is in no way a 'proof' of the Schrödinger equation ; it merely shows that if the de Broglie relationship is assumed, and if the motion of the electron is analogous to a system of standing waves, equation 32 is the type of wave equation to be expected. It was the genius of Schrödinger to arrive at equation 32 by intuition, and to justify this by solving the equation to give values for E in agreement with experiment. We shall discuss the way in which this was done in the next chapter.

THE INTERPRETATION OF ψ

The function $\psi(x, y, z)$ which occurs in the wave equation is equivalent to the function $f(x)$ which represented the amplitude in the case of the vibration of a stretched string. We can therefore refer to ψ as an amplitude function. Now we saw in the first two sections of this chapter that if the behaviour of electrons is represented by a wave equation, we can equate the square of the function representing the amplitude in the wave equation with either (a) the electron density, or (b) the probability that the electron will be found in a given volume element. We thus get a physical significance for the function $\psi^2(x, y, z)$, in that

$$\psi^2 \; . \; \mathrm{d}x \; . \; \mathrm{d}y \; . \; \mathrm{d}z \; (= \psi^2\,\mathrm{d}v)$$

measures the probability that the electron will be found in the volume element dv surrounding the point whose co-ordinates are (x, y, z). The other interpretation, that ψ^2 dv represents the electron density in the volume element dv, cannot be justified so rigorously as the probability interpretation, but it has proved to be very useful in practice. We have seen that difficulty arises when the concept of electron density is applied to systems containing only a single electron. What, for example, is the meaning of a statement that the electron density at a particular point is 0·2? We are tempted to recall JOHN DALTON's celebrated remark 'Thou

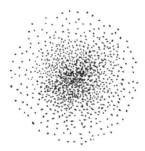

Figure 3.5. Charge cloud for a hydrogen electron

knowest no man can split an atom', and apply it to the electron. The concept may perhaps be made more plausible if we consider a hypothetical experiment. Suppose we could photograph the position of an electron at any particular instant; three-dimensional photography would enable us to assign coordinates x, y, z to the electron and to plot it as a point in a three-dimensional diagram. Now the electron is in rapid motion, and a photograph taken a fraction of a second later would reveal the electron in a new position. Several million photographs would produce an array of dots in our diagram which would resemble a cloud, dense in regions where a large number of points are crowded together, diffuse in regions where there are very few points. The dense regions would be those where there is a high probability of finding an electron. We thus get a link between particle density and probability. *Figure 3.5* represents a section through such a charge cloud for the electron in its state of lowest energy in the hydrogen atom. Although a few points will be at a considerable distance from the origin, it is possible to construct a surface which will enclose a large proportion of the points, say

31

95 per cent. We shall see in the next chapter that analogous electron charge cloud distributions, of different shapes, can be drawn for other energy states of the hydrogen atom electron.

REFERENCES

[1] DE BROGLIE, L. *Phil. Mag.* 47 (1924) 446
[2] DAVISSON, C., and GERMER, L. H. *Phys. Rev.* 30 (1927) 705
[3] THOMSON, G. P. *Proc. R. Soc.* A 117 (1928) 600
[4] SCHRÖDINGER, E. *Ann. Phys., Lpz.* 79 (1926) 361, 489.

THE HYDROGEN ATOM

We shall assume that the motion of the single electron in the hydrogen atom can be represented by the wave equation

$$\nabla^2 \psi + \frac{8\pi^2 \mu}{h^2} (E - V) \psi = 0 \qquad \ldots (33)$$

Here ψ is the wave function, h Planck's constant, E the total energy and V the potential energy. One small change from equation 32 may be noted; we have replaced the electron mass m by the 'reduced mass' μ, as defined on page 15. So far we have used a Cartesian system of coordinates, and have taken the nucleus as origin, defining the position of the electron by coordinates x, y, and z in the usual way. We shall find, however, that the mathematical handling of the wave equation becomes much simpler if we use spherical polar coordinates instead of Cartesian ones. Incidentally, the use of these coordinates is analogous to the method used to specify the position of a point on the earth's surface; the position is defined by the radius of the earth and by two angles, the latitude and the longitude. The simplification that such a transformation brings about can be illustrated by reference to the equation for a sphere; in Cartesian coordinates this is

$$x^2 + y^2 + z^2 = \text{constant}$$

whereas in spherical polar coordinates we get the simpler equation $r = \text{constant}$. The relationship between the Cartesian coordinates x, y, z and the polar coordinates r, θ, ϕ is shown in *Figure 4.1*; simple geometry gives

$$x = r \sin \theta \cos \phi$$
$$y = r \sin \theta \sin \phi$$
$$z = r \cos \theta \qquad \ldots (34)$$

If we now transform equation 33, using the relationships of equation 34, and putting $V = - e^2/r$ (*i.e.* the potential energy of the electron at a distance r from the nucleus), we get

$$\frac{1}{r^2} \cdot \frac{\partial}{\partial r}\left(r^2 \frac{\partial \psi}{\partial r}\right) + \frac{1}{r^2 \sin \theta} \cdot \frac{\partial}{\partial \theta}\left(\sin \theta \frac{\partial \psi}{\partial \theta}\right)$$

$$+ \frac{1}{r^2 \sin^2 \theta} \cdot \frac{\partial^2 \psi}{\partial \phi^2} + \frac{8\pi^2 \mu}{h^2}\left(E + \frac{e^2}{r}\right)\psi = 0 \qquad \ldots (35)$$

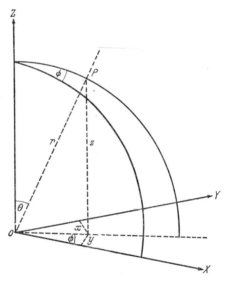

Figure 4.1. Relationship between Cartesian coordinates x, y, z and spherical polar coordinates r, θ, ϕ

This is a straightforward, though somewhat lengthy piece of mathematical manipulation and we merely state the result. The wave function, which is a function of three variables r, θ, and ϕ, can now be separated* into the product of three functions $R(r)$, $\Theta(\theta)$, and $\Phi(\phi)$ where R is a function of r only, Θ of θ only and Φ of ϕ only. Thus

$$\psi(r, \theta, \phi) = R(r) \cdot \Theta(\theta) \cdot \Phi(\phi)$$

If we substitute $R \cdot \Theta \cdot \Phi$ for ψ in equation 35 and then multiply through by $\dfrac{r^2 \sin^2 \theta}{R \cdot \Theta \cdot \Phi}$ we get the expression

* This can only be done for wave functions that do not vary with time.

$$\frac{\sin^2\theta}{R}\cdot\frac{d}{dr}\left(r^2\frac{dR}{dr}\right) + \frac{\sin\theta}{\Theta}\cdot\frac{d}{d\theta}\left(\sin\theta\frac{d\Theta}{d\theta}\right)$$

$$+\frac{1}{\Phi}\cdot\frac{d^2\Phi}{d\phi^2} + r^2\sin^2\theta\cdot\frac{8\pi^2\mu}{h^2}\left(E+\frac{e^2}{r}\right) = 0 \qquad \ldots(36)$$

Here, the third term, $\frac{1}{\Phi}\cdot\frac{d^2\Phi}{d\phi^2}$, is a function only of ϕ, and does not change its value when θ and r change, whereas the remaining terms are functions of θ and r only. Now the sum of all the terms in equation 36 can be zero for all values of ϕ only if $\frac{1}{\Phi}\cdot\frac{d^2\Phi}{d\phi^2}$ is constant, and we shall represent this constant by $-m^2$. This may seem a curious and arbitrary constant to choose, but the reason for selecting $-m^2$ is that it enables us to write down a very simple expression for the solution of the differential equation containing ϕ. This equation is

$$\frac{1}{\Phi}\cdot\frac{d^2\Phi}{d\phi^2} + m^2 = 0 \qquad \ldots(37)$$

Substitution of the equation 37 in equation 36 and dividing through by $\sin^2\theta$ gives us an equation in R and Θ,

$$\frac{1}{R}\cdot\frac{d}{dr}\left(r^2\frac{dR}{dr}\right) + \frac{1}{\Theta\sin\theta}\cdot\frac{d}{d\theta}\left(\sin\theta\frac{d\Theta}{d\theta}\right)$$

$$-\frac{m^2}{\sin^2\theta} + \frac{r^2\,8\pi^2\mu}{h^2}\left(E+\frac{e^2}{r}\right) = 0 \qquad \ldots(38)$$

Here the second and third terms are independent of r, and the first and last terms independent of θ; we can therefore separate these terms and put each equal to a constant. Thus if we put the terms involving r equal to β we get

$$\frac{1}{R}\cdot\frac{d}{dr}\left(r^2\frac{dR}{dr}\right) + \frac{r^2 8\pi^2\mu}{h^2}\left(E+\frac{e^2}{r}\right) = \beta \qquad \ldots(39)$$

The terms in θ must now be put equal to $-\beta$, in order to satisfy equation 38, giving

$$\frac{1}{\Theta\sin\theta}\cdot\frac{d}{d\theta}\left(\sin\theta\frac{d\Theta}{d\theta}\right) - \frac{m^2}{\sin^2\theta} = -\beta \qquad \ldots(40)$$

SOLUTION OF THE Φ EQUATION

We have now reduced our original differential equation to three very much simpler equations, but before we proceed to outline the method of solution we must recall the discussion in Chapter 3 (page 27). There will, in general, be an infinite number of solutions, and we have to choose those which are physically appropriate to an electron in a hydrogen atom. Just as in the case of the stretched string (page 28), the allowed wave functions for the electron in a hydrogen atom, whether expressed as ψ, R, Θ, or Φ, must be finite, continuous, and single-valued, if we are to equate their squares with the probability of finding the electron at a particular point. In addition, the wave function must be 'normalized'. Thus the electron in a hydrogen atom must be somewhere in space so that there is unit probability, *i.e.* certainty of finding it in a volume comprising the whole of space, and since $\psi^2 \cdot dv$ represents the probability that the electron is in the volume element dv, the wave functions for the electron in hydrogen must satisfy

$$\int \psi^2 \, dv = 1$$

where the integration is taken over all space. Wave functions which satisfy this condition are said to be normalized. Now it can readily be shown that

$$\Phi = a \sin m\phi \qquad \ldots (41)$$

satisfies equation 37; Φ is finite, since $a \sin \phi$ can never exceed the value a, and is continuous, being a sine function. Φ will only be single valued, however, if m is an integer, because if m is not an integer, an increase in ϕ by 2π will not bring us back to the same value of Φ. Thus equation 41 will provide acceptable wave functions if $m = 0$ or ± 1, ± 2, $\ldots$ *etc.*: m is called the magnetic quantum number, and we see that it appears as a direct consequence of the association of a wave function appropriate to a standing wave system with the motion of an electron in an atom. The condition that Φ should be normalized is

$$\int_0^{2\pi} \Phi^2 \, d\phi = 1; \ i.e. \ \int_0^{2\pi} a^2 \sin^2 m\phi \, d\phi = 1$$

This gives $a = 1/\sqrt{\pi}$ and so the solution of the Φ equation can finally be written:

$$\Phi = \frac{1}{\sqrt{\pi}} \cdot \sin m\phi \qquad \ldots (42)$$

where $m = 0, \pm 1, \pm 2, \ldots$ *etc.* The same argument can be used to show that

$$\Phi = \frac{1}{\sqrt{\pi}} \cdot \cos m\phi$$

is also an alternative solution of equation 37, and linear combinations of these two solutions will also be solutions of equation 37.

SOLUTIONS OF THE Θ AND R EQUATIONS

The solution of the Θ and R equations is a much more complicated matter, and we can here only state the results without proof. In the first place it can be shown that acceptable solutions of the Θ equation are only obtained if

$$\beta = l(l + 1)$$

l can have the values $|m|, |m| + 1, |m| + 2, \ldots$ *etc.*, where m is the magnetic quantum number introduced in the solution of the Φ equation; l is called the 'azimuthal quantum number' and, as we saw in Chapter 2, it is related to the k quantum number of the old quantum theory by $l = k - 1$. When we come to the solution of the R equation we find that for negative values of E satisfactory solutions require the introduction of a third quantum number; this is n, the 'principal quantum number', which can take the values $l + 1, l + 2, l + 3, \ldots$ *etc.* If E is positive there are no restrictions. These allowed values for the three quantum numbers thus involved in the solution of the Schrödinger equation for hydrogen can be written in a more familiar way if we start by giving n the values $1, 2, 3, 4, \ldots$ *etc.* The other quantum numbers then take the values

$$l = 0, 1, 2, \ldots (n - 1)$$

and

$$m = -l, -l + 1, \ldots -1, 0, +1,$$
$$\ldots +(l - 1), +(l)$$

We can now write the complete wave function for the hydrogen electron as

$$\psi_{nlm}(r, \theta, \phi) = R_{nl}(r) \cdot \Theta_{lm}(\theta) \cdot \Phi_m(\phi)$$

Such a wave function, describing a single electron, is called an 'orbital'. Each orbital will have its associated quantum numbers n, l, and m, and it is convenient to have a method of distinguishing orbitals with different quantum numbers; the nomenclature already introduced in Chapter 2 is therefore taken over into the new quantum theory. Orbitals with azimuthal quantum numbers $l = 0, 1, 2, 3, \ldots$ are called $s, p, d, f, \ldots$ respectively, and the value of the principal quantum number n is written in front of this letter. Thus $3d$ refers to an orbital for which $n = 3$ and $l = 2$.

THE PHYSICAL SIGNIFICANCE OF THE s ORBITALS

The solution of the wave equation for $n = 1$, $l = 0$, $m = 0$ gives the following expression for the $1s$ orbital

$$\psi_{1s} = \frac{1}{\sqrt{\pi}} \cdot \left(\frac{1}{a_0}\right)^{0/2} \cdot e^{-r/a_0}$$

The corresponding expression for the $2s$ orbital ($n = 2$, $l = 0$, $m = 0$) is

$$\psi_{2s} = \frac{1}{4\sqrt{2\pi}} \cdot \left(\frac{1}{a_0}\right)^{3/2} \cdot \left(2 - \frac{r}{a_0}\right) \cdot e^{-r/2a_0}$$

where a_0 has been written for $h^2/4\pi^2 \mu e^2$. It will be noted that the expressions for these orbitals are functions of r and independent of θ and ϕ. They are therefore spherically symmetrical, so that we get the same value of ψ at a distance r from the nucleus no matter what values we give to θ and ϕ. Now we saw in Chapter 3 that we can get a physical interpretation of the wave function for a single electron in an atom by equating $\psi^2 dv$ with the probability of finding the electron in a volume element dv; thus a graph of ψ^2 against r will show the 'probability distribution', from which we can obtain the chance of finding the electron in a volume element dv at a distance r from the nucleus. *Figure 4.2* shows this probability distribution for the $1s$ and $2s$ orbitals of hydrogen. Let us first confine our attention to the $1s$ orbital. It will be seen that there is a high probability of finding the electron in a volume element close to the nucleus, but it is also important to realize that there is a finite (though very small) chance of finding the electron at an almost infinite distance away from the nucleus. There is an alternative, and often more useful, way of interpreting the wave function, which is to consider the probability that the electron is in a spherical 'shell', of radius r and thickness dr, centred on the nucleus. The volume of such a shell is $4\pi r^2 dr$, and the probability that the electron is in this shell is given by the product of the probability that the electron is in the volume element dv at a distance *r*, and the volume of the shell $4\pi r^2 dr$. *Figure 4.3* shows the function $4\pi r^2 \psi^2$, which is called the 'radial probability distribution',

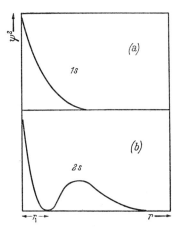

Figure 4.2. Probability distribution for the 1*s* and 2*s* electrons of hydrogen

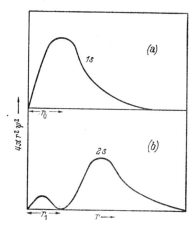

Figure 4.3. (*a*) Radial probability distribution for the 1*s* electron of hydrogen; (*b*) radial probability distribution for the 2*s* electron of hydrogen

plotted against *r* for the hydrogen 1*s* and 2*s* orbitals. We see in *Figure 4.3(a)* that there is a maximum probability of finding the electron in a shell of radius $r = r_0$, for although the probability of finding the electron in a volume dv is greatest close to the nucleus, the volume of a spherical shell close to the nucleus is very small.

The numerical value of r_0 for the $1s$ orbital of hydrogen can be calculated: it turns out to be $0 \cdot 529\text{Å}$, which is also the value obtained on the old quantum theory for the radius of the orbit with principal quantum number $n = 1$. The important difference between the orbital and the Bohr orbit should be noted. Whereas the Bohr theory confined the electron to a fixed orbit, the new quantum theory describes the electron by an orbital, which gives the electron a maximum probability of being at a distance from the

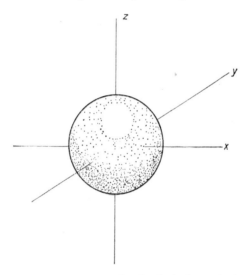

Figure 4.4. Charge cloud surface for the hydrogen $1s$ electron

nucleus equal to the Bohr radius, but it also gives the electron a finite probability of being either closer to the nucleus, or further away from it. Now although there is a finite chance of finding the electron at a very great distance from the nucleus, this chance is very small; the probability distribution function, ψ^2, falls off very rapidly as r increases. We can draw a sphere around the nucleus of such a radius that there is a high probability that the spherical surface will enclose the electron, and we can select an arbitrary value, say 95 per cent, for this probability, thus getting an approximate, but very useful, geometrical representation of the electron distribution in the $1s$ state of hydrogen. *Figure 4.4* is a drawing of such distribution; (the dots are used merely to give a three-dimensional effect; they should not be confused with dots used to denote electron positions, as in *Figures 3.5* and *4.5*). A section through a diameter of the sphere will be a circle, and we shall, in general, use the circle as a two-dimensional method of illustrating the spherical 1s distribution. The argument can now be extended to cover other s type orbitals of higher principal quantum number. *Figure 4.2 (b)* shows that the probability function ψ^2 for the 2s orbital falls steeply to zero at $r = r_1$, passes through a shallow maximum, and then falls away to infinity. There is, therefore, a spherical nodal surface of radius r_1 surrounding the nucleus. We can draw a diagram to represent the charge cloud similar to that used in *Figure 3.5*, page 31,

Figure 4.5. Charge cloud for the hydrogen 2s electron

to describe a 1s electron; *Figure 4.5* shows the result to be expected if we could get a series of photographs of the position of the electron at different instants. A comparison of *Figures 4.3 (a)* and *(b)* shows that an increase in the principal quantum number corresponds to an increase in the space effectively filled by the charge distribution. It must be emphasized, however, that there is still only *one* electron in the atom, and the average electron density in the 2s state must, therefore, be less than in the 1s state. The argument can be extended to orbitals of higher values of n. When $n = 2$ there is one nodal surface; when $n = 3$ there are two such surfaces and, in general, there are $(n - 1)$ nodal surfaces in an s orbital of principal quantum number n.

THE PHYSICAL SIGNIFICANCE OF THE p ORBITALS

There are no p orbitals when $n = 1$ since the only value that the azimuthal quantum number l can take is zero. For $n = 2$ and $l = 1$, there are three p orbitals, $2p_{+1}$, $2p_0$ and $2p_{-1}$, corresponding to m values of 1, 0 and -1, respectively. We shall find it convenient, however, to use linear combinations of these solutions, and the mathematical expressions for these linear combinations for the hydrogen atom are given by:

$$\psi(2p_x) = \mathcal{N} \cdot R(r) \cdot \sin \theta \cos \phi$$
$$\psi(2p_y) = \mathcal{N} \cdot R(r) \cdot \sin \theta \sin \phi$$
$$\psi(2p_z) = \mathcal{N} \cdot R(r) \cdot \cos \theta$$

where $R(r)$ is the radial function (*i.e.* a function of the radius r), and $\mathcal{N}$ a normalizing constant. These $2p$ wave functions for the hydrogen atom electron represent states of equal energy, and such states are said to be 'degenerate'.

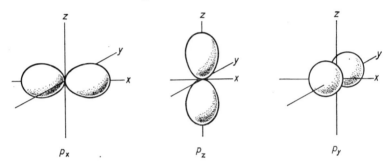

p_x $\qquad$ p_z $\qquad$ p_y

Figure 4.6. Θ^2 (θ). Φ^2 (ϕ) surfaces for the $2p_x$, $2p_y$, and $2p_z$ electrons

The fact that these wave functions are functions of θ and ϕ as well as of r means that it is not easy to get a simple physical model of the p orbitals; a four-dimensional diagram would be needed to show the variation of ψ with r, θ and ϕ simultaneously.

We have, therefore, to separate our wave function into the product of radial and angular functions, $R(r) \cdot \Theta(\theta) \cdot \Phi(\phi)$, and draw separate diagrams for the variation of R^2 against r, Θ^2 against θ, and Φ^2 against ϕ. The product of these functions $R^2 \cdot \Theta^2 \cdot \Phi^2$ will then give us the probability of finding the electron in a volume element dv surrounding a point whose coordinates are r, θ, ϕ. It is possible, however, to draw a diagram which will show simultaneously the variation of the function $\Theta^2 \cdot \Phi^2$ with θ and ϕ. The resulting surface for the $2p_z$ electron is shown in *Figure 4.6*. This shows that the angular part of the probability distribution is concentrated along the z axis. (Reference to *Figure 4.1* will show that the z axis corresponds to $\theta = 0$, $\phi = 0$, the y axis to $\theta = \pi/2$, $\phi = \pi/2$, and the x axis to $\theta = \pi/2$, $\phi = 0$.) It is also important to note that Θ^2 . Φ^2 is zero in the xy plane, since $\cos^2 \theta$ is zero for points in this plane. Such a plane, in which the amplitude of the wave function is zero, is called a 'nodal plane'. We can draw diagrams showing the Θ^2 . Φ^2 distribution for the $2p_x$ and $2p_y$ electrons, and these are also shown in *Figure 4.6* They have the same shape as the $2p_z$ distribution, but different orientations; the angular probability distribution of the $2p_x$ function is concentrated in the direction of the x axis, with the yz plane a

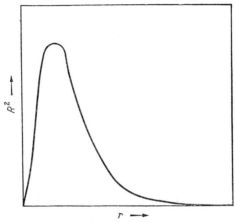

Figure 4.7. Radial wave function for a $2p$ electron

nodal plane, and the $2p_y$ distribution is concentrated along the y axis with xz as the nodal plane. The probability of finding the electron in a volume dv (r, θ, ϕ) is then obtained by multiplying Θ^2 . Φ^2 by R^2. *Figure 4.7* shows R^2 plotted against r for the $2p$ electrons (R is independent of θ and ϕ and so the radial function is the same for $2p_x$, $2p_y$, and $2p_z$). We see that R^2 is zero at the origin, increases to a maximum, then falls off very rapidly as r increases to infinity. This rapid decrease in the value of R^2 as r increases means that we can again get a reasonable, though approximate, physical picture of the probability distribution by drawing surfaces corresponding to those shown in *Figure 4.6,* so that there is, say, a 95 per cent probability of finding the electron inside the volume so enclosed. The variation of electron density within this surface can be conveniently represented by contours—lines along which the electron density is constant. *Figure 4.8* shows such contours for a section through the $\Theta^2(\theta) \cdot \Phi^2(\phi)$ surface for a p_x electron.

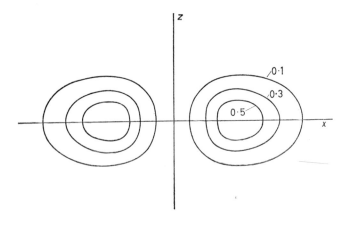

Figure 4.8. Electron density contours for a $2p_x$ orbital

THE PHYSICAL SIGNIFICANCE OF THE d ORBITALS

We have seen that for the principal quantum number $n = 2$, there are four orbitals, one s and three p. When $n = 3$ there are nine orbitals, and in general, for every value of n there are n^2 orbitals. The nine orbitals for $n = 3$ are obtained as shown in *Table 4.1*.

Table 4.1

Type of orbital	Quantum numbers $n \quad l \quad m$			Number of orbitals
$3s$	3	0	0	1
$3p$	3	1	$0, \pm 1$	3
$3d$	3	2	$0, \pm 1, \pm 2$	5

Here again it is convenient to use linear combinations of the solutions corresponding to d_0, $d_{\pm 1}$ and $d_{\pm 2}$; the mathematical expressions for these five independent linear combinations for the $3d$ orbitals of hydrogen are given below, where we have written $R(r)$ to represent the part of the expression dependent on r, while $N_1, N_2, \ldots$ *etc.* represent normalizing constants.

44

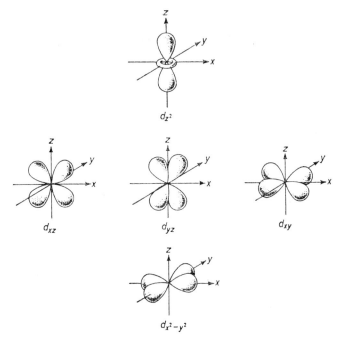

d_{z^2}

d_{xz} d_{yz} d_{xy}

$d_{x^2-y^2}$

Figure 4.9. Θ^2 (θ) . Φ^2 (ϕ) surfaces for the $3d$ electrons

$$\psi_{3d(i)} = d_{z^2} = N_1 R(r)(3 \cos^2 \theta - 1)$$
$$\psi_{3d(ii)} = d_{xz} = N_2 R(r)(\sin \theta \cos \theta \cos \phi)$$
$$\psi_{3d(iii)} = d_{yz} = N_3 R(r) (\sin \theta \cos \theta \sin \phi)$$
$$\psi_{3d(iv)} = d_{x^2-y^2} = N_4 R(r)(\sin^2 \theta \cos 2\phi)$$
$$\psi_{3d(v)} = d_{xy} = N_5 R(r)(\sin^2 \theta \sin 2\phi)$$

Figure 4.9 gives the polar diagrams showing the Θ^2 . Φ^2 patterns. Cartesian subscripts (xy, yz, x^2, $x^2 - y^2$, *etc.*) are used, since they represent the angular part of the wave function. Thus, transforming to polar coordinates by means of equations 34 on page 33 we get

$$xz = r^2(\sin \theta \cos \theta \cos \phi)$$

so that the angular part of $\psi_{3d(ii)}$ can be written d_{xz}. Similarly,

$$x^2 - y^2 = r^2(\sin^2 \theta \cos^2 \phi - \sin^2 \theta \sin^2 \phi)$$
$$= r^2 \sin^2 \theta \cos 2 \phi$$

so that the angular part of $\psi_{3d(iv)}$ can be written as $d_{x^2-y^2}$, and so on for the other orbitals. The d_{z^2} function should logically be written d_{3z^2-1}, but the shorter form d_{z^2} is always used.

The four orbitals labelled d_{xy}, d_{yz}, d_{xz} and $d_{x^2-y^2}$ have identical shapes but different orientations with respect to a set of Cartesian

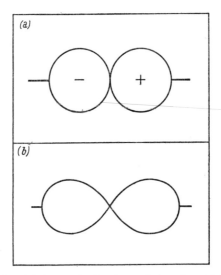

Figure 4.10. (a) *Cross-section* showing variation of ψ with θ and ϕ; (b) *cross-section* showing variation of ψ^2 with θ and ϕ

axes; the axes of the d_{xy} and $d_{x^2-y^2}$ lobes lie in the xy plane with maximum extension either along the x and y axes $(d_{x^2-y^2})$ or along lines making 45° with these axes (d_{xy}). The axes of the d_{yz} lobes are in the yz plane with maximum extension along lines making 45° with the y and z axes, while the axes of the d_{xz} lobes are in the xz plane, with maximum extension along lines making 45° with the x and z axes. The d_{z^2} orbital, however, has a different shape from the other four since, in addition to a lobe with maximum extension along the z axis, there is also a 'tyre' or 'collar' surrounding the z axis with maximum electron density in the xy plane. It is not possible to select five independent d orbitals with the same shape, but it should be noted that the d_{z^2} orbital can be written as a linear combination of two orbitals $(d_{y^2-z^2}, d_{z^2-x^2})$ which do have the same shape as the four discussed above.

We must now try to clarify a matter that is often confused by students. The surfaces described in the two preceding sections, and illustrated in *Figures 4.4, 4.6* and *4.9*, have been surfaces showing the function $\Theta^2(\theta) \cdot \Phi^2(\phi)$ and, therefore, when multiplied by $R^2(r)$ have the physical significance of probability distributions. We shall often find it very convenient, however, when we are discussing directed valencies, to draw surfaces to describe orbitals. Now the orbital is the function ψ, not ψ^2, and so the orbital surfaces are not necessarily identical with the ψ^2 surfaces previously drawn. The surfaces for the s orbitals are the same shape as the ψ^2 surfaces—both being spherical—but those for the p orbitals differ. *Figure 4.10* shows the angular dependence of ψ and ψ^2 for the p orbitals. The ψ surface is that of two spheres in contact; we get the same general properties as in the surfaces describing the probability distribution, the same orientation of the p_x, p_y, and p_z orbitals along the x, y, and z axes, and the same nodal planes, but the shapes are different. The probability patterns are the physically significant ones, and the orbital patterns the mathematically useful ones, as we shall see in later chapters. There is one further point of some importance. The function $\Theta \cdot \Phi$ may be positive or negative according to the values of θ and ϕ, and in the p orbitals one lobe is positive and the other negative as shown in the diagram. There is no such differentiation in the 'probability' distributions, since ψ^2 must be positive.

ENERGY LEVELS

It now remains to discuss the energy of the electron in the different orbitals of the hydrogen atom. The energy E occurs in equation 39, the radial equation. We saw that satisfactory wave functions could only be obtained, for negative values of E, if a quantum condition was imposed; for each value of n there is a corresponding energy. It can be shown that this energy is given by

$$E_n = - \frac{2\pi^2 \mu e^4}{h^2 n^2}$$

an expression identical with that obtained on the Bohr theory (page 13). This is a very important result, for it can be tested experimentally, and there is good agreement between the energies calculated by this equation and those determined by spectroscopic measurements. (If E is positive there are no quantum restrictions. This corresponds to situations where the electron is far removed from the influence of the nucleus, and moving with varying kinetic energy which may have any value. It is the existence of these non-quantized energy levels that leads to the production of a continuous spectrum.) The normal or 'ground' state of the hydrogen atom is that for which $n = 1$, which gives the state of lowest energy. Higher values of n correspond to states of higher energy, the so-called excited states. (Remember that E is negative, and that the energies are measured from a zero which corresponds to the state where the electron is at infinite distance from the nucleus.) We can plot our energy values on a vertical scale and get a series of energy levels just as we did in Chapter 2 (*Figure 2.2*). For the hydrogen atom, the treatment outlined above leads to the conclusion that all the orbitals having the same principal quantum number have the same energy; *e.g.* the $3s$, $3p$, and $3d$ orbitals are on the same energy level. A more refined treatment, introducing relativity theory, predicts slight energy differences between the levels, and explains the fine structure in the hydrogen spectrum.

5

QUANTUM THEORY AND THE
PERIODIC CLASSIFICATION

THE WAVE EQUATION FOR MANY-ELECTRON ATOMS

DIFFICULTIES at once arise when we try to apply the methods of the preceding chapter to atoms containing more than one electron. In these atoms, any particular electron is attracted to the oppositely charged nucleus, but it is also repelled by the other electrons present, and the expression for the potential energy of such an electron involves terms containing inter-electronic distances. It is the presence of these terms in the wave equations that makes an exact solution impossible, and approximation methods have to be used. One widely used method is that developed by D. R. and W. H. HARTREE. In effect, they reduce the problem to that of a single electron moving in a spherically symmetrical field of force provided by the nucleus and the averaged effect of all the other electrons. The complete wave function for the atom is then written as the product of a number of one-electron wave functions. Some typical results of these lengthy calculations are shown in *Figure 5.1*, where the radial distribution function $4\pi r^2 \psi^2$ is plotted against r for the sodium ion, Na^+. The shaded portion shows the electron distribution in Na^+, and the broken line marked 3s shows the distribution curve for the outermost electron in the neutral sodium atom. The diagram shows that in the Na^+ ion the electron charge cloud effectively forms two concentric shells quite close to the nucleus, and there is evidently a strong resemblance to the orbits of the Bohr theory. Solution of the wave equation again introduces the quantum numbers n and l, which retain the significance they had in the case of hydrogen; thus the 3s distribution curve is that of an electron with principal quantum number $n = 3$ and azimuthal quantum number $l = 0$. Similar distribution curves have been worked out by Hartree and his pupils for a large number of ions and neutral atoms. They all reveal a characteristic arrangement of electrons in concentric shells, the inner electrons in general being close to the nucleus.

ENERGY LEVELS

The energies corresponding to different orbitals in many-electron atoms differ in two ways from the corresponding energies in the hydrogen atom. In the first place, the increased nuclear charge will make it more difficult to remove an electron from the atom, and secondly, the screening of the nucleus by inner electrons will affect the energy levels. *Figure 5.1* shows that the distribution curve for the 3*s* electron overlaps very considerably with the inner

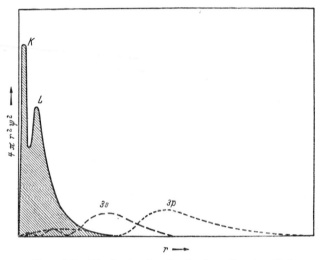

Figure 5.1. Distribution function for the sodium ion, Na+

electron distribution given by the shaded portion of the diagram; this means, in effect, that the 3*s* electron is under the influence of an almost unscreened nuclear charge. The dotted line in *Figure 5.1* gives the distribution curve for a 3*p* orbital, where the overlap with the inner electrons is very small and the nucleus is effectively screened. It will therefore require less energy to remove a 3*p* electron from an atom than a 3*s* one, since the effective nuclear charge is smaller in the former case. We saw in Chapter 3 that the energy levels can be obtained from atomic spectra observations, and *Figure 5.2* shows two sets of experimentally determined energy levels, the left-hand column giving the relative positions for atoms of low atomic number (less than 20) and the right-hand column giving relative positions for heavy elements. It should be noted that, for orbitals of the same principal quantum number, the order of increase of energy is $s < p < d$. The diagram also shows that, for elements of atomic number less than 20, the energy of the 3*d* orbital is higher than the 4*s* orbital, but that for elements of atomic number greater than 20 this order is reversed. Similar changes

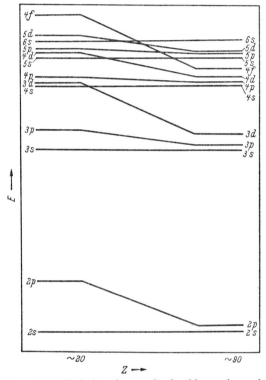

Figure 5.2. Variation of energy levels with atomic number

occur in the relative positions of the 4*f* and 5*d* orbitals before and after the atomic number 57. The importance of these energy differences will be discussed in the last section of this chapter.

ELECTRON SPIN

In Chapter 2, page 18, we introduced the idea of the spinning electron to explain the doublet structure of the alkali metal spectra. So far we have not introduced the electron spin into the new quantum theory and, in fact, it does not appear in the Schrödinger method. It does appear, however, in the solution of a considerably more abstract wave theory developed by P. A. M. DIRAC (1928), which combines quantum theory and relativity theory. We have to add to the three quantum numbers already introduced, n, l, m, the spin quantum number m_s which is allowed only two values, $\pm \frac{1}{2}$. The quantum numbers n, l, and m retain the same significance that they had for hydrogen; n largely determines the energy associated with the orbital, l indicates the geometrical shape of the orbital, m is associated with different orientations of the orbital with reference to some defined direction. It is not possible to give an adequate simple physical significance to m_s. Thus we may have, for example, when $n = 2$, one spherically symmetrical $2s$ orbital, and three dumb-bell shaped p orbitals, $2p_x$, $2p_y$, and $2p_z$ orientated at right-angles to each other. When $n = 3$ there will be, in addition, the possibility of five d type orbitals. The shape of these orbitals will resemble those described for the hydrogen atom electron, but their size will be greater. (They differ very considerably from the orbits of the Bohr theory; thus a Bohr orbit with $l = 0$ is a very narrow elongated ellipse (*see* page 16), whereas on the new quantum theory $l = 0$ corresponds to a spherical s orbital.)

Figure 5.1 shows that the inner electrons are held close to the nucleus so that in the alkali metals the size and shape of the charge cloud is mainly determined by the outermost electron. The approximation method, which treats the atom as a single electron system, works very well in these cases, and the quantum numbers assigned in the way just described have an exact meaning. In other atoms the approximation is not so good, particularly when there may be several electrons with the same principal quantum number; however, it is still possible to allocate four quantum numbers to each electron in the atom.

THE EXCLUSION PRINCIPLE

We can now consider the way in which the electrons in an atom distribute themselves among the various orbitals. The first guiding principle is that electrons will tend to go into the orbitals of lowest energy. Thus the single electron in a hydrogen atom will go into a $1s$ orbital; the electronic configuration in hydrogen is therefore described briefly as $1s^1$, where the superscript shows that there is a single electron in an orbital of principal quantum number $n = 1$ and azimuthal quantum number $l = 0$. It might at first sight be thought that in other atoms all the electrons will also crowd into the $1s$ orbital, since this has the lowest energy. This is indeed the case in helium, where the configuration is summarized by $1s^2$.

However, the interpretation of the atomic spectrum of helium requires these two electrons to have opposite spins; if they had parallel spins the appearance of the spectrum would be quite different. The lithium spectrum can be interpreted if we treat it as a one-electron system. The observed frequencies can be related to a number of transitions of the outer electron between orbitals of different energy, and principal quantum number $n = 2$ and azimuthal quantum number $l = 0$ can be assigned without difficulty to this outer electron. We therefore assign the configuration $1s^2 2s^1$ to the lithium atom in its stable state, noting that apparently the $1s$ orbital can accommodate only two electrons. Here we have an example of the working of what is probably the most important principle in theoretical chemistry. It was first explicitly stated by the Swiss physicist PAULI in 1925, and it is now referred to as the *Pauli Exclusion Principle.* He pointed out that the correlation of experimental observations on line spectra with the quantum theory required the introduction of a postulate, *viz.*

> that in any system, whether a single atom or a molecule, no two electrons could be assigned the same set of four quantum numbers.

Thus, the two electrons in helium, possessing the three quantum numbers n, l, and m, must have different values of m_s, and no other electrons can enter the $1s$ orbital. No proof of the exclusion principle exists but, so far, nothing in nature has appeared to contradict it, and its acceptance brings a new order into the whole of science.

It follows from the exclusion principle that, if there are a number of electrons with parallel spins in an atom, these electrons must be in different orbitals. They tend to be as far away from each other as possible, and the direction of the bonds formed by such electrons is thus largely determined by the operation of the exclusion principle. *Table 5.1* shows the maximum number of electrons that can be accommodated in the different orbitals when the exclusion principle is taken into account.

When $n = 1$ there is only a single s orbital into which two electrons of opposite spin can go, but when $n = 2$ we can have both s and p type orbitals. The three p orbitals, $2p_x$, $2p_y$, and $2p_z$ (related to the three possible m values, ± 1, 0), can each hold two electrons with opposite spins, making a total of eight electrons for the full set of orbitals with this principal quantum number. It can be shown, in the same way, that for $n = 3$ and $n = 4$ the corresponding maximum numbers of electrons are 18 and 32, respectively. The

Table 5.1. The Distribution of Electrons in Orbitals

n	l	m	m_s	Total number of electrons
1	0	0	$\pm\frac{1}{2}$	2
2	0	0	$\pm\frac{1}{2}$	2 ⎫
		± 1	$\pm\frac{1}{2}, \pm\frac{1}{2}$	⎬ 8
	1	0	$\pm\frac{1}{2}$	6 ⎭
3	0	0	$\pm\frac{1}{2}$	2 ⎫
	1	± 1	$\pm\frac{1}{2}, \pm\frac{1}{2}$	
		0	$\pm\frac{1}{2}$	6
		± 2	$\pm\frac{1}{2}, \pm\frac{1}{2}$	⎬ 18
	2	± 1	$\pm\frac{1}{2}, \pm\frac{1}{2}$	10
		0	$\pm\frac{1}{2}$	⎭

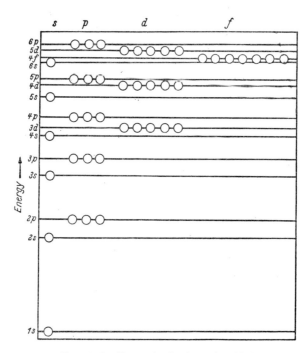

Figure 5.3. Energy levels of atomic orbitals

orbitals associated with the various energy levels are shown in diagrammatic form in *Figure 5.3* (no geometrical significance is to be attached to the use of circles; they simply represent orbitals into which two electrons can go).

IONIZATION ENERGIES

The distribution of electrons in the various orbitals, determined in an empirical way by the operation of the exclusion principle, is supported experimentally by the analysis of atomic spectra and also by the measurement of the so-called ionization potentials or ionization energies. The ionization potential is defined as the energy needed to displace the most easily removable electron from its equilibrium position in the atom to infinity. This energy can be determined by electrical methods, which usually give a value quoted in electron volts—hence the term ionization potential. Chemists, however, preferring to quote energies in kilogram calories per mole, tend nowadays to speak of ionization energies. Many electron atoms have a series of ionization energies. The *first* ionization energy is the energy needed to move one electron from a neutral atom to infinity; *e.g.* for $He \rightarrow He^+$, $I_1 = 567 \cdot 0$ kcal/mol. The *second* ionization energy is defined as the energy needed to remove an electron from a singly-charged ion to infinity; *e.g.* $He^+ \rightarrow He^{2+}$, $I_2 = 1,255$ kcal/mol. I_2 is greater than I_1 since the negatively charged electron has to be dragged away from an ion with a net positive charge.

Table 5.2. Ionization Energies

Atom	First ionization energy, kcal/mol.
H	313·6
He	567·0
Li	124·3
Be	215·0
B	191·4
C	259·8
N	335·4
O	314·1
F	401·8
Ne	497·3
Na	118·5
Mg	176·3
Al	138·0
Si	187·9
P	243·4
S	238·9
Cl	300·1
Ar	363·5

Table 5.2 gives numerical values for the first ionization energies of elements in the first two rows of the Periodic Table, and *Figure 5.4* plots these values against atomic number.

We see that the energy needed to remove an electron from a lithium atom is very much less than that needed to remove an electron from helium. The configuration Li, $1s^2 2s^1$ explains this. Ionization of a lithium atom requires the removal of the electron in a $2s$ orbital which is effectively screened from the nuclear charge by the 'inner' electrons in the $1s$ orbital. The electrons in a helium atom, configuration $1s^2$ are not effectively screened, and so very much more energy is needed to remove one of them.

The next break in the first ionization energy plot comes at beryllium. We see that it takes more energy to remove an electron from this atom than it does to remove one from its neighbours lithium and boron. The configuration $1s^2 2s^2$ is evidently more stable than that for lithium. The exclusion principle would require the electron configuration for boron to be $1s^2 2s^2 2p^1$ (see *Figure 5.3*), and the experimental value for the first ionization energy shows that an electron in a $2p$ orbital is more effectively screened from the nuclear charge than one in the $2s$ orbital which is paired with another $2s$ electron of opposite spin.

THE 'BUILDING-UP' ('AUFBAU') PRINCIPLE
AND THE PERIODIC CLASSIFICATION

We are now in a position to make use of the Exclusion Principle and the experimental information available from atomic spectra and ionization energy measurements, to assign electron configurations to atoms, taking them in order of increasing atomic number. In doing so, we shall also use another important generalization, called the 'Building-Up' or 'Aufbau' principle. This states that in determining the configuration of an atom of atomic number Z, we first of all write down the configuration of the atom of atomic number $Z - 1$; we then have to consider only the allocation of quantum numbers to the one additional electron, assuming this electron to go into the available orbital of lowest energy. So far we have assigned H $1s^1$, He $1s^2$, Li $1s^2 2s^1$, Be $1s^2 2s^2$ and B $1s^2 2s^2 2p^1$. The filling up of the three $2p$ orbitals continues in this way from carbon to neon. In each case the inner electron distribution remains unchanged as $1s^2$—the helium configuration—and it will be convenient in future to write (He) to represent this arrangement. We therefore arrive at the following electron configurations: carbon $(He)2s^22p^2$, nitrogen $(He)2s^22p^3$, oxygen $(He)2s^22p^4$, fluorine $(He)2s^22p^5$ and neon $(He)2s^22p^6$. These configurations are consistent with the observed atomic spectra and ionization energies of the elements. *Figure 5.4* clearly shows that the neon configuration represents a particularly stable arrangement from which it is difficult to remove an electron, and the very much smaller value for the ionization energy of sodium indicates that the electron

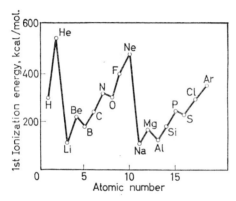

Figure 5.4. The first ionization energies of the elements in the first two short periods

removed is in an orbital well screened from the nuclear charge, *i.e.* $3s$. Low ionization energy values such as those shown for the alkali metals can be correlated with particularly stable electron configurations for the ion which remains when the electron has been removed, *e.g.* Li^+1s^2, $Na^+1s^22s^22p^6$, etc. The low value for boron reflects the stability of the configuration B^+, $1s^22s^2$, while the low value for oxygen compared with its adjacent elements is perhaps related to the configuration of the O^+ ion, $1s^22s^22p_x^12p_y^12p_z^1$ in which there is one electron in each of the three p orbitals. We shall see later that analogous configurations, where the available d or f orbitals are just half-filled, are more stable than arrangements containing one electron more or less than this number. Study of atomic spectra also leads to another very useful empirical principle, usually known as Hund's rule: that when a set of p (or d or f) orbitals is being filled with electrons, the distribution is such that the electrons retain parallel spins as far as possible. Thus the two $2p$ electrons in carbon will be arranged $2p_x^1 2p_y^1$, since the exclusion principle will prevent them from retaining parallel

57

spins if both are in a single p orbital. The electronic configuration $(He)2s^22p_x^2$ is known, but it corresponds to an 'excited' state of the carbon atom, with an energy 29 kcal greater than the stable state $(He)2s^22p_x^12p_y^1$. The stable nitrogen configuration, when written out in full, will therefore be $(He)2s^22p_x^12p_y^12p_z^1$, the configuration $(He)2s^22p_x^22p_y^1$ referring to an excited state 55 kcal greater in energy. Pairing of electrons with opposed spins will have to take place in oxygen, and the structure is written $(He)2s^22p_x^22p_y^12p_z^1$. For neon we get a completely filled set of p orbitals in which there are no 'unpaired' electrons. Reference to *Figure 5.3* shows that the next orbitals to be occupied will be the $3s$, $3p$, and $4s$ ones although it must be noted that for elements of atomic number greater than 20, the relative positions of the $4s$ and $3d$ energy levels is changed. This change will be discussed in the next section. We can now write down the configurations of the atoms from sodium to calcium inclusive, noting that the inner electron arrangement in each case is that of neon which will be written for conciseness as (Ne). Thus we get sodium $(Ne)3s^1$, magnesium $(Ne)3s^2$, aluminium $(Ne)3s^23p^1$, . . . , and so on up to argon $(Ne)3s^23p^6$. Potassium and calcium have the inner electron arrangement of argon, and their configurations therefore become $(Ar)4s^1$ and $(Ar)4s^2$, respectively. We can now see how the chemical resemblances between different elements, so brilliantly expressed by MENDELÉEFF in the periodic system, are paralleled by resemblances in electronic structure. In particular, we observe that the alkali metals have one unpaired electron in the outermost orbital, that the alkaline earth metals have two electrons in the outer orbital and, especially, that the noble gases are characterized by completely filled s and p type orbitals.

It has sometimes been stated that the periodic system could have been derived from observations on atomic spectra and application of the quantum theory. It seems at least doubtful that this could have been achieved. The men who developed the quantum theory and the methods of interpreting atomic spectra were brought up in a tradition in which the Mendeléeff classification was part of the scientific climate of the age; the sorting out of the complexities of atomic spectra would scarcely have been possible without the conscious, or unconscious, utilization of the periodic system. The quantum theory, with many triumphs to its credit, has not yet surpassed the achievement of Mendeléeff in predicting accurately the properties of the then undiscovered elements scandium, gallium, and germanium. However, the quantum theory does resolve some of the defects of the old periodic system. (The anomalous position of argon and potassium, tellurium and iodine, and nickel and cobalt had already been rectified by the use of an arrangement based on atomic number rather than atomic weight, before the development of the new quantum theory.) The outstanding defects were the apparently arbitrary position of the noble gas elements, the difficulty of accommodating the rare earth elements, and the existence in the same group of not very similar elements. The last defect was recognized by separation into A and B sub-group elements, but this hardly provided an explanation of the difference between them. We have already seen that the quantum theory provides a logical position for the noble gases in the periodic system—they are atoms in which the s and p orbitals are completely filled. The resolution of the remaining defects will be discussed in the next section.

TRANSITION SERIES

The next element to consider is scandium, atomic number 21. Analysis of its atomic spectrum shows quite clearly that the electronic structure cannot be similar to that of boron or aluminium, *i.e.* it cannot be, what might possibly have been expected, $(Ar)4s^2 4p^1$. Further study of the spectrum reveals that for scandium and for atoms of greater atomic number, the $3d$ orbital, so far unoccupied, is now more stable than the $4p$ one, and thus the electronic configuration of scandium is $(Ar)3d^1 4s^2$. *Figure 5.3* shows the position of the five $3d$ orbitals with respect to the neighbouring $4s$ and $4p$ orbitals.

Beginning with scandium, we get a series of ten elements (ending at zinc) which are characterized, in general, by two outer electrons in a $4s$ orbital, and from one to ten electrons in an inner $3d$ orbital. The filling up of the d orbitals is not quite regular; there is not much difference in energy between the $3d$ and $4s$ orbitals, and chromium and copper find their respective configurations of $3d^5 4s^1$ and $3d^{10} 4s^1$ to be more stable than the expected $3d^4 4s^2$ and $3d^9 4s^2$; a particularly stable state seems, in fact, to be formed either when the five d orbitals are completely filled with ten electrons, or when there is one electron in each of the five. The same general principles which operated in the first two periods still apply here: in a series of equivalent orbitals (*e.g.* the five $3d$), electrons arrange themselves as far as possible with parallel spins. Thus, the stable (or 'ground') state of manganese $(3d^5 4s^2)$ contains five unpaired electrons in the $3d$ orbitals. Manganese readily forms a stable divalent Mn^{2+} ion $(3d^5)$, and iron $(3d^6 4s^2)$ a stable tervalent

Table 5.3. *Electron Configurations for the Ground States of the Elements (numbers in parentheses are uncertain)*

Element	Atomic number	K	L		M			N				O				P				Q
		1s	2s	2p	3s	3p	3d	4s	4p	4d	4f	5s	5p	5d	5f	6s	6p	6d	6f	7s
H	1	1																		
He	2	2																		
Li	3	2	1																	
Be	4	2	2																	
B	5	2	2	1																
C	6	2	2	2																
N	7	2	2	3																
O	8	2	2	4																
F	9	2	2	5																
Ne	10	2	2	6																
Na	11	2	2	6	1															
Mg	12	2	2	6	2															
Al	13	2	2	6	2	1														
Si	14	2	2	6	2	2														
P	15	2	2	6	2	3														
S	16	2	2	6	2	4														
Cl	17	2	2	6	2	5														
Ar	18	2	2	6	2	6														
K	19	2	2	6	2	6		1												
Ca	20	2	2	6	2	6		2												
Sc	21	2	2	6	2	6	1	2												
Ti	22	2	2	6	2	6	2	2												
V	23	2	2	6	2	6	3	2												
Cr	24	2	2	6	2	6	5	1												
Mn	25	2	2	6	2	6	5	2												
Fe	26	2	2	6	2	6	6	2												
Co	27	2	2	6	2	6	7	2												
Ni	28	2	2	6	2	6	8	2												
Cu	29	2	2	6	2	6	10	1												
Zn	30	2	2	6	2	6	10	2												
Ga	31	2	2	6	2	6	10	2	1											
Ge	32	2	2	6	2	6	10	2	2											
As	33	2	2	6	2	6	10	2	3											
Se	34	2	2	6	2	6	10	2	4											
Br	35	2	2	6	2	6	10	2	5											
Kr	36	2	2	6	2	6	10	2	6											
Rb	37	2	2	6	2	6	10	2	6			1								
Sr	38	2	2	6	2	6	10	2	6			2								
Y	39	2	2	6	2	6	10	2	6	1		2								
Zr	40	2	2	6	2	6	10	2	6	2		2								
Nb	41	2	2	6	2	6	10	2	6	4		1								
Mo	42	2	2	6	2	6	10	2	6	5		1								
Tc	43	2	2	6	2	6	10	2	6	(5)		(2)								
Ru	44	2	2	6	2	6	10	2	6	7		1								
Rh	45	2	2	6	2	6	10	2	6	8		1								
Pd	46	2	2	6	2	6	10	2	6	10										
Ag	47	2	2	6	2	6	10	2	6	10		1								
Cd	48	2	2	6	2	6	10	2	6	10		2								
In	49	2	2	6	2	6	10	2	6	10		2	1							
Sn	50	2	2	6	2	6	10	2	6	10		2	2							

Table 5.3 (continued)

Ele-ment	Atomic number	K 1s	L 2s	2p	M 3s	3p	3d	N 4s	4p	4d	4f	O 5s	5p	5d	5f	P 6s	6p	6d	6f	Q 7s
Sb	51	2	2	6	2	6	10	2	6	10		2	3							
Te	52	2	2	6	2	6	10	2	6	10		2	4							
I	53	2	2	6	2	6	10	2	6	10		2	5							
Xe	54	2	2	6	2	6	10	2	6	10		2	6							
Cs	55	2	2	6	2	6	10	2	6	10		2	6			1				
Ba	56	2	2	6	2	6	10	2	6	10		2	6			2				
La	57	2	2	6	2	6	10	2	6	10		2	6	1		2				
Ce	58	2	2	6	2	6	10	2	6	10	(2)	2	6			(2)				
Pr	59	2	2	6	2	6	10	2	6	10	(3)	2	6			(2)				
Nd	60	2	2	2	2	6	10	2	6	10	(4)	2	6			(2)				
Pm	61	2	2	6	2	6	10	2	6	10	(5)	2	6			(2)				
Sm	62	2	2	6	2	6	10	2	6	10	6	2	6			2				
Eu	63	2	2	6	2	6	10	2	6	10	7	2	6			2				
Gd	64	2	2	6	2	6	10	2	6	10	(7)	2	6	(1)		2				
Tb	65	2	2	6	2	6	10	2	6	10	(8)	2	6	(1)		2				
Dy	66	2	2	6	2	6	10	2	6	10	(10)	2	6			(2)				
Ho	67	2	2	6	2	6	10	2	6	10	(11)	2	6			(2)				
Er	68	2	2	6	2	6	10	2	6	10	(12)	2	6			(2)				
Tm	69	2	2	6	2	6	10	2	6	10	13	2	6			2				
Yb	70	2	2	6	2	6	10	2	6	10	14	2	6			2				
Lu	71	2	2	6	2	6	10	2	6	10	14	2	6	1		2				
Hf	72	2	2	6	2	6	10	2	6	10	14	2	6	2		2				
Ta	73	2	2	6	2	6	10	2	6	10	14	2	6	3		2				
W	74	2	2	6	2	6	10	2	6	10	14	2	6	4		2				
Re	75	2	2	6	2	6	10	2	6	10	14	2	6	5		2				
Os	76	2	2	6	2	6	10	2	6	10	14	2	6	6		2				
Ir	77	2	2	6	2	6	10	2	6	10	14	2	6	7		2				
Pt	78	2	2	6	2	6	10	2	6	10	14	2	6	9		1				
Au	79	2	2	6	2	6	10	2	6	10	14	2	6	10		1				
Hg	80	2	2	6	2	6	10	2	6	10	14	2	6	10		2				
Tl	81	2	2	6	2	6	10	2	6	10	14	2	6	10		2	1			
Pb	82	2	2	6	2	6	10	2	6	10	14	2	6	10		2	2			
Bi	83	2	2	6	2	6	10	2	6	10	14	2	6	10		2	3			
Po	84	2	2	6	2	6	10	2	6	10	14	2	6	10		2	4			
At	85	2	2	6	2	6	10	2	6	10	14	2	6	10		2	5			
Rn	86	2	2	6	2	6	10	2	6	10	14	2	6	10		2	6			
Fr	87	2	2	6	2	6	10	2	6	10	14	2	6	10		2	6			1
Ra	88	2	2	6	2	6	10	2	6	10	14	2	6	10		2	6			2
Ac	89	2	2	6	2	6	10	2	6	10	14	2	6	10		2	6	(1)		(2)
Th	90	2	2	6	2	6	10	2	6	10	14	2	6	10		2	6	(2)		(2)
Pa	91	2	2	6	2	6	10	2	6	10	14	2	6	10	(2)	2	6	(1)		(2)
U	92	2	2	6	2	6	10	2	6	10	14	2	6	10	(3)	2	6	(1)		(2)
Np	93	2	2	6	2	6	10	2	6	10	14	2	6	10	(5)	2	6			(2)
Pu	94	2	2	6	2	6	10	2	6	10	14	2	6	10	(6)	2	6			(2)
Am	95	2	2	6	2	6	10	2	6	10	14	2	6	10	(7)	2	6			(2)
Cm	96	2	2	6	2	6	10	2	6	10	14	2	6	10	(7)	2	6	(1)		(2)
Bk	97	2	2	6	2	6	10	2	6	10	14	2	6	10	(8)	2	6	(1)		2
Cf	98	2	2	6	2	6	10	2	6	10	14	2	6	10	(10)	2	6			2
Es	99	2	2	6	2	6	10	2	6	10	14	2	6	10	11	2	6			2
Fm	100	2	2	6	2	6	10	2	6	10	14	2	6	10	12	2	6			2
Md	101	2	2	6	2	6	10	2	6	10	14	2	6	10	13	2	6			2
Lw	102	2	2	6	2	6	10	2	6	10	14	2	6	10	14	2	6			2
No	103	2	2	6	2	6	10	2	6	10	14	2	6	10	14	2	6	1		2

61

4

Fe^{3+} ion $(3d^5)$; the manganic ion Mn^{3+} $(3d^4)$ and the ferrous ion Fe^{2+} $(3d^6)$ are less stable. It is nowadays the custom to refer to this group of ten elements $(Sc \rightarrow Zn)$ as the 'First Transition Series', a term formerly restricted to the group VIII triad (iron, cobalt, and nickel). There are considerable chemical similarities between neighbouring members of the series, a fact related to the similar configurations in the outermost orbital. The presence of an incompletely filled set of d orbitals also confers characteristic properties, such as variable valency and the existence of coloured, paramagnetic ions; these will be discussed elsewhere in this book.

Zinc has an electronic structure with completely filled $3d$ and $4s$ orbitals, but there are still the three unoccupied $4p$ orbitals to consider. These become filled in the elements from gallium to krypton; gallium has the configuration $(Ar)3d^{10}4s^24p^1$, and krypton $(Ar)3d^{10}4s^24p^6$. Elements in the second long period from rubidium to xenon have configurations analogous to those just discussed. A second series of transition elements starts with yttrium which has the arrangement $(Kr)4d^15s^2$, but the filling up of the $4d$ orbitals does not occur in an entirely regular manner. Details of the individual configurations are given in *Table 5.3*. The $4d$ orbitals are completely occupied in cadmium, and the $5p$ orbitals then fill up, giving, when fully occupied, the noble gas structure of xenon.

In the next long period, caesium, barium, and lanthanum have the expected structures $(Xe)6s^1$, $(Xe)6s^2$, and $(Xe)5d^16s^2$. We should then expect to find that lanthanum was the first member of a third transition series of ten elements; in fact, this is not the case. The atomic spectrum of cerium does not closely resemble that of titanium or zirconium, elements in corresponding positions in the first two transition series. Reference to the energy levels of *Figure 5.2*, page 51, for an atom of high atomic number $(Ce = 58)$ shows that the energy of the $4f$ orbitals is lower than that of the $5d$ ones; in fact, cerium takes the configuration* $4f^25s^25p^65d^06s^2$, and we get a series of fifteen elements from lanthanum to lutecium where the main electronic differences are in the number of electrons occupying an inner set of seven $4f$ orbitals. Chemical resemblances between these elements are very marked, since the two outer orbitals remain almost unchanged. These elements (formerly called rare earths, but now generally referred to as the lanthanides), thus fall naturally into place in an arrangement of the elements based on electronic configuration. The filling up of the $4f$ orbitals as we go from lanthanum to lutecium is not entirely regular.

---

* Here we quote only the configuration in the outer orbitals.

Configurations in which the seven f orbitals are completely occupied $(4f^{14}5s^25p^6)$, or completely empty $(4f^05s^25p^6)$, or with a single electron in each of the seven orbitals $(4f^75s^25p^6)$, seem particularly stable, and the variable valency of some of the lanthanide elements can be explained in terms of the tendency to form these stable structures. After lutecium we reach hafnium which, with the configuration $4f^{14}5s^25p^65d^26s^2$, has a normal transition series structure, resembling titanium and zirconium. This transition series is complete at mercury, and in the later elements, the $6p$ orbitals are occupied until we get the inert gas radon, with the configuration $4f^{14}5s^25p^65d^{10}6s^26p^6$. We come finally to the incomplete period, the first two members of which are the radioactive alkali metal francium with configuration $(Rn)7s^1$, and radium $(Rn)7s^2$. The configuration of the remaining elements cannot as yet be given with complete certainty. Some of these elements have been obtained in only microscopic amounts, and in any case the interpretation of their very complex spectra is far from complete. It seems, however, to be generally accepted that, possibly in thorium, and almost certainly in protoactinium, $5f$ orbitals are being occupied. Configurations of individual actinide elements are given in *Table 5.3*. Actinium, with the configuration $5f^06s^26p^66d^17s^2$, forms the first member of the 'actinide' series. It strongly resembles lanthanum, the first member of the 'Lanthanide' series, in its chemical and physical properties. Both elements form insoluble fluorides, hydroxides, oxalates and phosphates, and the crystal structures of corresponding compounds are very similar.

Before leaving this section we must emphasize that the electron configurations here described refer to isolated atoms or ions in the gaseous state. The configuration may well be different if the ion is in a solid structure or in solution.

II

QUANTUM THEORY OF VALENCY

6

HISTORICAL INTRODUCTION

THE idea that chemical forces are electrical in nature can be traced back to the beginning of the nineteenth century. In 1800, NICHOLSON and CARLISLE decomposed water into hydrogen and oxygen by passing an electric current through it, and in the next few years many other electrolytic decompositions were reported. Perhaps the outstanding examples were Sir HUMPHRY DAVY's isolation of sodium and potassium from caustic soda and caustic potash, respectively, in 1807. Davy, indeed, suggested that the forces governing chemical combination were electrical in nature, the electrification of the combining particles being produced by contact. Electrochemical ideas were also developed by BERZELIUS in his Dualistic Theory (1812). He assumed that every atom possessed two electric poles of opposite sign, electropositive atoms having the positive pole in excess, electronegative atoms the negative pole in excess. The combination of an element with oxygen might produce a basic oxide (e.g. CuO) with a residual positive polarity, or an acidic oxide (e.g. SO_3) with a residual negative polarity. These oxides could combine because of the attraction of the residual opposite charges: thus

$$\overset{+}{CuO} + \overset{-}{SO_3} = CuSO_4$$

The theory, when subsequently applied to organic chemistry, could not, however, explain the fact that the substitution of negative chlorine for positive hydrogen in many organic molecules produced comparatively little change in chemical properties (cf. CH_3COOH and CCl_3COOH). Moreover, it could not account for FARADAY's Laws of Electrolysis.

Electrochemical theories of chemical combination were then neglected for many years; the advances in theoretical organic chemistry associated with the names of LAURENT, GERHARDT, FRANKLAND, WILLIAMSON, and KEKULÉ did not refer specifically to the electrical nature of combining forces in molecules. The electrical theory was, however, restated by HELMHOLTZ in his Faraday Lecture of 1881, when he said ' . . . the very mightiest among the chemical forces are of electric origin. The atoms cling to their electric charges, and opposite charges cling to each other'.

ARRHENIUS, in 1884 and 1887, published a theory of 'electrolytic dissociation' in which he proposed the idea that salts (*e.g.* sodium chloride) in dilute aqueous solution were dissociated into positive (*e.g.* sodium) ions and negative (*e.g.* chloride) ions. This theory of easily dissociated or 'ionizable' atoms was used by WERNER (1891), who discussed the constitution of compounds of the type $CoCl_3 . 6NH_3$ in terms of 'principal' and 'auxiliary' valencies. The auxiliary valencies were exerted in the co-ordination of a number of atoms, molecules or radicals to the central metal atom, in an 'inner sphere of combination', and the principal valencies represented the attachment of ionizable atoms or groups in an 'outer sphere of combination'. Thus Werner represented $CoCl_3 \cdot 6NH_3$ as

where the dotted lines represent the 'auxiliary' bonds and the full lines the attachment of the ionizable atoms. Werner's theory was subsequently of great importance in the study of isomerism in inorganic compounds, and in other ways; we shall refer to it more fully in Chapter 12.

J. J. THOMSON's (1897) recognition of the negatively charged electron as a constituent of all atoms, and his measurements of the ratio of its charge to its mass, provided a new stimulus to the electro-chemical theories. He realized that the chemical properties of elements depended in some way on the arrangement of their electrons, and he suggested that electropositive atoms were those which could achieve a stable electronic state by losing one or two electrons, whereas electronegative atoms achieved stable states by acquiring one or more electrons.

Sir WILLIAM RAMSAY's Presidential Address to the Chemical Society of London in 1908 stressed the role of the electron in valency. Thus he said ' . . . they (the electrons) serve as the "bonds of union" between atom and atom'. He also suggested that in molecules such as NaCl and Cl_2, the electron might form a 'cushion' between the two atoms. Ramsay, and other chemists of this period, believed that the maximum number of electrons involved in compound formation was eight; thus nitrogen, with five available electrons, could acquire three more by combining with three hydrogen atoms to form ammonia, NH_3, whereas NH_4 could only be obtained by removing one electron and forming a positively charged NH_4^+ ion.

ABEGG (1904) had postulated that elements had two valencies, a 'normal' and a 'contravalency'. These were of opposite polarity, and the total sum of the two valencies was always eight. J. NEWTON FRIEND (1908) distinguished three types of valency—negative, positive, and residual—where the negative valency of an atom was defined as being equal to the number of electrons with which it could combine, and the positive valency was related to the loss of electrons from the atom. His use of the term 'residual valency' is of great interest in the light of later developments, for he suggested that when an element exerted its residual valency, it simultaneously parted with, and received, an electron. The hydrogen molecule was represented as $H \rightleftharpoons H$, the arrows indicating the directions in which the electrons were transferred.

J. J. Thomson made a further important contribution to valency theory in 1914. He emphasized the difference between polar molecules (*e.g.* NaCl) and non-polar ones (*e.g.* most organic substances), and observed that the electropositive valency of an element was equal to the number of electrons that could easily be separated from it, while the negative valency was equal to the difference between eight and the positive valency.

The development of the Rutherford theory of the nuclear atom, and Bohr's work on the hydrogen atom structure (*see* page 12) paved the way for a more comprehensive 'electronic theory of valency'. The foundations of this modern theory were laid in two independent publications by W. KOSSEL and by G. N. LEWIS in 1916. Kossel, who was mainly concerned with polar (ionizable) molecules, pointed out that in the periodic system, a noble gas element always separates an alkali metal and a halogen, and that the formation of a negative ion by the halogen atom, and a positive ion by the alkali metal atom, would give each of these atoms the structure of a noble gas. The noble gases were assumed to possess a particularly stable configuration of eight electrons in an outer shell (two electrons in the case of helium). No compounds of these elements were known at that time—indeed until very recently they were usually called the 'inert' gases. A number of compounds of xenon have now been made, but it seems unlikely that stable compounds of helium, neon and argon can be formed, and these elements might well still be called inert gases. The stable ions in a compound such as NaCl are, on this view, held together by electrostatic attraction and form what is now called an 'electrovalent' bond or link. The electrovalency of the ion is defined as the number of unit charges on the ion; thus magnesium has a positive electrovalency of two, while chlorine has a negative electrovalency of one. We shall discuss the formation and properties of electrovalent (now usually called ionic) bonds in some detail in Chapter 10.

Lewis discussed atomic structure in terms of a positively charged 'kernel' (*i.e.* the nucleus plus the 'inner' electrons), and an outer shell that could contain up to eight electrons. He assumed that these outer electrons were arranged at the corners of a cube surrounding the kernel; thus, the single electron in the outer shell of sodium would occupy one corner of the cube, whereas all eight corners would be occupied by the electrons in the outer shell of an inert gas. This octet of electrons represented a particularly stable electronic arrangement, and Lewis suggested that, when atoms were linked by chemical bonds, they achieved this stable octet of electrons in their outer shells. Atoms such as sodium and chlorine could achieve an outer octet by the transfer of an electron from sodium to chlorine, forming Na^+ and Cl^- ions, respectively. This was essentially the mechanism proposed by Kossel, but Lewis proposed a second mechanism to account for the formation of non-polar molecules. Here, there was no transfer of electrons from one atom to another (and thus no ion formation), but the bond resulted from the sharing of a pair of electrons, each atom contributing one electron to the pair.

This theory was considerably extended by LANGMUIR (1919), although he abandoned the idea of the stationary cubical arrangement of the outer electrons; he introduced the term 'covalent bond' to describe the Lewis 'electron-pair' bond or link. We can illustrate the Lewis-Langmuir theory by considering the chlorine molecule, Cl_2. Chlorine, with the electronic configuration (using the modern notation introduced in Chapter 5), $(Ne)3s^23p^5$, is one electron short of the inert gas configuration of argon, (Ne) $3s^2 3p^6$, and the formation of the stable diatomic chlorine molecule, Cl_2, results from the sharing of electrons by the two chlorine atoms. If we represent the outer electrons of one chlorine atom by dots, and those of the other by crosses, we can write

$$:\overset{\cdot}{\underset{\cdot\cdot}{C}}l\cdot \quad + \quad \overset{\times\times}{\underset{\times\times}{C}}l\times \quad = \quad :\overset{\cdot\cdot}{\underset{\cdot\cdot}{C}}l : \overset{\times\times}{\underset{\times\times}{C}}l\times$$

where Cl represents the chlorine nucleus and the inner electrons. The atoms linked in this way need not be identical; thus, in carbon tetrachloride, each of the four outer electrons of the carbon pairs with an electron from a chlorine atom to form four covalent bonds. The Lewis-Langmuir theory would represent the structure as

$$
\begin{array}{c}
{}^{xx}_{x}\overset{x}{C}l{}^{x}_{x} \\[2pt]
{}^{x}_{x}\overset{xx}{C}l{}^{x}_{.}\overset{.}{C}{}^{.}_{.}\overset{xx}{C}l{}^{x}_{x} \\[2pt]
{}^{x}_{x}\overset{x}{C}l{}^{x}_{xx}
\end{array}
$$

Double and triple bonds are considered to involve the sharing of four and six electrons, respectively, so that the structures of ethylene and acetylene would be written

$$
\begin{array}{c}
H_{x} \quad {}_{x}H \\
{}^{.}_{x}C \; {}^{x}_{x} \; C{}^{.}_{x} \\
H^{.} \quad {}^{.}H
\end{array}
\qquad \text{and} \qquad
H \overset{.}{x} C \; {}^{xxx}_{xxx} \; C \overset{x}{.} H
$$

In these molecules (Cl_2, CCl_4, C_2H_4, and C_2H_2) the electron pair for each single bond is provided by the two combining atoms, each contributing one electron. PERKINS (1921) postulated a related type of link in which both electrons for the electron-pair bond come from one only of the two combining atoms. An example is provided in the combination of ammonia with boron trimethyl, $B(CH_3)_3$. Ammonia may be written

$$
\begin{array}{c}
H \\[2pt]
H \; {}^{x}_{.}\overset{..}{N}{}^{.}_{.} \\[2pt]
H
\end{array}
$$

each hydrogen atom pairing its electron with a nitrogen electron to form three electron-pair bonds; the two unused outer electrons on the nitrogen atom form a so-called 'lone pair'. In boron trimethyl, however,

$$
\begin{array}{c}
Me \\[2pt]
Me \; {}^{x}_{.}\overset{.}{B}{}^{.}_{.} \\[2pt]
Me
\end{array}
\qquad \text{(where Me represents the methyl radical } H \overset{.}{x} C \overset{..}{x} H \text{)}
$$

there are only six electrons around the boron atom, since the boron atom itself only has three outer electrons, which it uses to form three electron-pair bonds with the carbon atoms of the three methyl groups, as shown above. A compound $H_3N \rightarrow BMe_3$ can now be formed. The nitrogen atom is the 'donor' and the boron atom the 'acceptor' in this operation; the arrow indicates the relationship donor → acceptor. It should be emphasized that 'donation' is a special case of electron 'sharing', and no complete transfer of electrons takes place; nevertheless, the nitrogen atom has, in effect, 'lost' one electron (it now only shares two electrons both of which it had entirely to itself), and the boron atom has, so to say, 'gained' an electron. The formation of this type of bond thus involves a charge displacement, producing what is called an electric 'dipole' in the molecule. (We refer again to this term in Chapter 12.) Nitrogen acquires a 'formal' positive charge, and boron a 'formal' negative charge in $H_3N \rightarrow BMe_3$. Apart from this charge displacement the bond, once formed, does not differ in any way from the normal covalent bond of the Lewis-Langmuir theory.

The Lewis-Langmuir electron-pair or covalent bond is now often referred to as the homopolar bond, whereas the complete transfer of electrons, resulting in ion formation (e.g. Na^+, Mg^{2+}, Cl^-), gives rise to the electrovalent, or ionic, or heteropolar bond, by the attraction of opposite charges. Bonds formed by the Perkins mechanism were originally called 'co-ordinate' links, but the terms 'donor' or 'dative' bond have also been used. The production of formal charges on the atoms linked in this way is emphasized by the term 'semi-polar' bond, which implies that the bond has something of the character of the polar (i.e. ionic) link; a term recently introduced by PALMER (1944)—the co-ionic bond—also indicates that the bond has characteristics of both the covalent and the ionic bond. In bond diagrams, the covalent bond may be represented either by the colon, as in Cl : Cl, or, more usually, by the dash, Cl—Cl. Co-ordinate links may be represented, as we saw in the $H_3N \rightarrow BMe_3$ example, by a single arrow showing the direction of the electronic charge displacement. An alternative method is to write $H_3\overset{+}{N} - \overset{-}{B}Me_3$, where $+$ and $-$ are used to indicate the formal charges. The different names given above are often a source of confusion to beginners, so they are collected together for clarity as follows:

Polar = Heteropolar = Ionic = Electrovalent

Non-polar = Homopolar = Covalent

Semi-polar = Co-ordinate = Dative = Co-ionic = Donor-Acceptor

The first calculation of the energy, and thus the strength, of a covalent electron-pair bond was made in 1927 by HEITLER and

LONDON, who applied quantum theory to the problem of the structure of the hydrogen molecule. Schrödinger equations can be written down to describe the behaviour of electrons in molecules, but solutions of these equations can only be obtained by using approximation techniques. The Heitler-London treatment is called the electron-pair or valence-bond method; it has been improved and extended by SLATER and by PAULING. An alternative treatment, known as the molecular-orbital method, was developed at the same time (1927) by Burrau for the hydrogen molecule ion (H_2^+); this method was developed (ca. 1932) by MULLIKEN and by LENNARD-JONES and their pupils. Both methods are approximations, but in a particular case, one technique may be more appropriate than the other. We shall describe the molecular-orbital treatment in Chapter 7 and the valence-bond method in Chapter 8.

BIBLIOGRAPHY

A fuller account of the historical development of valency theory will be found in the following works:

NEWTON FRIEND, J. *The Theory of Valency*, Longmans, Green Co., London, 1909

PARTINGTON, J. R. *A Short History of Chemistry*, Macmillan Co., London, 1939

BERRY, A. J. *Modern Chemistry—Some Sketches of its Historical Development*, Cambridge University Press, London, 1946

7

THE MOLECULAR-ORBITAL METHOD

INTRODUCTION

THE molecular-orbital theory starts by considering a system in which the nuclei are in their equilibrium positions in the stable molecule, and discusses the way in which the electrons, associated in some way with all the nuclei, can be described by wave functions. The procedure, which is analogous to that used for atomic systems, may be summarized as follows:

1. Each electron is placed in a molecular energy state or molecular orbital, which is described by appropriate quantum numbers. The molecular orbital will be polynuclear, that is to say it is associated with all the atomic nuclei present in the molecule, and the wave function for the orbital will have the same physical significance that it had for single atoms—i.e. $\psi^2 dv$ is proportional to the probability of finding the electron in a given volume of space dv.

2. Each molecular wave function corresponds to a definite energy value, and the sum of the individual energies of electrons in the molecular orbitals, after correction for interaction, represents the total energy of the molecule.

3. The 'Aufbau' principle may be applied. The electrons are fed into the available molecular orbitals one at a time, the lowest energy molecular orbital being filled first. Each electron has a spin, and by the Pauli Exclusion Principle (page 52) each molecular orbital can accommodate a maximum of two electrons, providing their spins are opposed.

The molecular orbitals may be obtained by several methods, but we shall only discuss that known as the 'linear combination of atomic orbitals' (abbreviated to LCAO).

THE LCAO METHOD

Let us consider a simple homonuclear diatomic molecule, *i.e.* two identical atoms linked by an electron-pair bond. (Although the atoms are identical, it will be convenient to distinguish the two nuclei by writing the molecule as A—B).

Now, although the Schrödinger equation for a system containing more than one electron cannot, in general, be solved completely to give accurate wave functions describing the behaviour of electrons in the molecule, it is possible to make shrewd guesses at the form of these functions, with the help of chemical and mathematical intuition, and previous experience. In the molecule $A—B$ we can assume that in the immediate neighbourhood of atom A, an electron is acted upon only by the nucleus A and any associated electrons. The wave function ϕ_A which describes the electron in this situation corresponds to the atomic wave function we should get for an isolated atom A. In a similar way, another wave function ϕ_B will describe the behaviour of the electron when it is close to nucleus B. These are extreme cases, for the electron will be under the influence of both the nuclei A and B. We assume that a suitable approximate or 'trial' molecular orbital is a linear combination of the atomic orbitals ϕ_A and ϕ_B, i.e.

$$\psi = N\{c_A\phi_A + c_B\phi_B\}$$

where N is a normalizing constant, chosen to ensure that $\int \psi^2 dv$ taken over the whole of space is unity, and c_A and c_B are numerical coefficients. (We recall the discussion on page 25, where we emphasized that if functions ϕ_1 and ϕ_2 are solutions of a wave equation, the linear combination $(a\phi_1 + b\phi_2)$ is also a solution.)

The next step is to find values for the coefficients c_A and c_B which will give the most satisfactory 'trial' function, i.e. the one which is nearest to the true molecular orbital, and this can be done using a technique called the 'Variation Method'. We can write down a large number of approximate wave functions ψ by giving different values to c_A and c_B, and associated with each of these functions there will be a corresponding energy $E_1, E_2, \ldots E_n$. Any linear combination of ϕ_A and ϕ_B will give an associated energy lower than that corresponding to ϕ_A or ϕ_B separately, and the Variation Method states that the nearest approximation to the true wave function will be the combination giving the lowest energy. When applied to the molecule $A—B$ this means that E must be a minimum with respect to c_A and c_B, so that, in mathematical terms, we get two linear equations

$$\frac{\partial E}{\partial c_A} = 0 \text{ and } \frac{\partial E}{\partial c_B} = 0$$

which can be solved to give E and the ratio c_A/c_B. The equation from which the energy E can be obtained has two roots, one for which $c_A = c_B$, and one for which $c_A = -c_B$, and there are, therefore, two linear combinations,

$$\psi_+ = N c_A \{\phi_A + \phi_B\}$$

and

$$\psi_- = N c_A \{\phi_A - \phi_B\}$$

The function ψ_+ gives a charge cloud in which there is a build-up of electron density in the region between the nuclei and, therefore, a more effective screening of one nucleus from the other; this brings about the formation of a bond, since there is a decrease in energy compared with the energies of the separated atoms. The bond linking atoms A and B is thus described by the function ψ_+ which is called a 'bonding' molecular orbital. We say that this molecular orbital has been formed by the overlap of the two atomic orbitals represented by ϕ_A and ϕ_B.

A detailed examination of the expression for the energy associated with the function ψ_+ shows that effective combination of the component functions ϕ_A and ϕ_B only occurs if

(a) they represent states of similar energy,

(b) they overlap to a considerable extent, and

(c) they have the same symmetry with respect to the molecular axis A—B.

If these conditions are not satisfied, the coefficients c may have very small values, and the corresponding wave functions will make little contribution to the linear combination.

The function ψ_- is called an 'antibonding' orbital, since it represents a state of higher energy, in which electrons are displaced from the internuclear region.

The energy of the antibonding molecular orbital is greater than that of either atomic orbital, as *Figure 7.1* indicates, so that if the electron should find itself in the antibonding orbital, it would return if possible to a more stable atomic orbital of lower energy. This point will become increasingly significant when more complex molecules are discussed, in which electrons may occupy both bonding and antibonding molecular orbitals; the presence of an electron in an antibonding orbital introduces a factor opposing the formation of a stable molecule.

We now illustrate the molecular-orbital method by reference to the hydrogen molecule ion, H_2^+, in which two protons are held together by a single electron. This molecule can be detected spectroscopically when hydrogen under reduced pressure is subjected to an electric discharge, and it is found to have a bond length of 1·06 Å and a dissociation energy of 2·791 eV. The atomic orbitals ϕ_A and ϕ_B used in the LCAO procedure will be the hydrogen $1s$ orbitals (see page 38), and the linear combination of these two orbitals will produce two molecular orbitals, one bonding and one antibonding. (Usually, the combination of n atomic orbitals gives n molecular orbitals, half of which are bonding and half antibonding.) The single electron in the ground state of H_2^+ occupies the bonding orbital. This comparatively simple approximation method gives a calculated energy which has a mini-

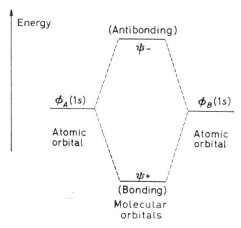

Figure 7.1. The relative energies of molecular orbitals and their constituent atomic orbitals

mum value at an internuclear separation of 1·32 Å, when the dissociation energy is 1·76 eV. Although these values are not very close to the experimental ones, it must be appreciated that it is something of an achievement that these results should be of the right order of magnitude. Calculated values for the internuclear distance and dissociation energy which agree exactly with experimental values have been obtained, but only by making very considerable modifications to the trial functions, with a corresponding elaboration of the computations involved. The bonding and antibonding molecular orbitals, and their relationship to the atomic $1s$ orbitals, are illustrated in Figure 7.2. The significance of such boundary surfaces for atomic orbitals was discussed in Chapter 4, and it may be worth pointing out yet again that these surfaces (or cross-sections of surfaces) merely represent the region in which the electron is most likely to be found. The maximum 'electron density' is normally well inside such surfaces. As we shall see shortly, this representation of atomic orbitals with positive and negative signs is helpful in choosing combinations of the correct symmetry.

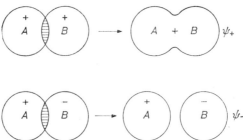

Figure 7.2. The combination of *s* atomic orbitals to form molecular orbitals

The wave functions ψ_+ and ψ_- are usually written as ψ_g and ψ_u in molecular-orbital theory. The subscripts *g* and *u* are used for the German words 'gerade' (even) and 'ungerade' (odd), and refer to important symmetry properties of the wave functions. An orbital is said to be symmetrical (gerade) if the wave function is unchanged when the coordinates *x, y, z* of the electron are replaced by $-x, -y, -z$, *i.e.* when any point in the orbital is reflected in the mid point or centre of symmetry; conversely, an orbital is antisymmetrical (ungerade) if the wave function changes sign when the *x, y, z,* coordinates are replaced by $-x, -y, -z$.

It is also convenient to introduce new symbols to show the type of atomic orbitals that have combined to produce bonding and antibonding molecular orbitals. Molecular orbitals formed by a linear combination of 1*s* atomic orbitals are called $\sigma 1s$ type orbitals, the bonding one being written $\sigma 1s$ and the antibonding one $\sigma^* 1s$. Combination of two 2*s* atomic orbitals will produce $\sigma 2s$ and $\sigma^* 2s$ molecular orbitals; these will be of higher energy than the $\sigma 1s$ orbitals since they are formed from atomic orbitals of higher energy. These orbitals are seen to be symmetrical about the molecular axis A—B; this is a characteristic feature of σ orbitals.

Some new problems appear when we discuss the combination of *p* type atomic orbitals to form molecular orbitals. Here we have to remember two facts about *p* orbitals: first, there are three equivalent *p* orbitals for a given principal quantum number, p_x, p_y, and p_z, which are set at right angles to each other, and secondly, the two halves of the *p* orbital have opposite mathematical signs;

one lobe is positive and the other negative. When the p_x orbitals combine (*cf. Figure 7.3*) we get bonding and antibonding molecular orbitals similar to the σs and $\sigma^* s$ orbitals; they are symmetrical about the molecular axis A—B (*i.e.* the x axis), and are denoted σp and $\sigma^* p$. The combination of p_y orbitals, however, produces

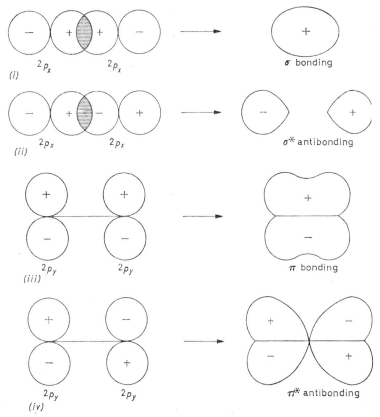

Figure 7.3. The combination of p_x atomic orbitals to form σ and σ^* molecular orbitals (*i, ii*); the combination of p_y orbitals to form π and π^* molecular orbitals (*iii, iv*)

molecular orbitals of very different shapes, since, in this case, both the positive and the negative lobes of the p orbital overlap. The bonding orbital consists of two electronic charge clouds concentrated into 'streamers', and the molecular axis A—B lies in a nodal plane perpendicular to the plane of the diagram. *Figure 7.3* shows a cross-section through these 'streamers'. (The positive and negative portions of the orbital may be visualized, somewhat crudely, as the two halves of a meringue, where the cream forms the nodal plane.) The shape of the antibonding orbital reveals the characteristic withdrawal of the electrons from the internuclear region, so that the repulsion between the nuclei is greater; this orbital has a higher energy than that of the bonding orbital. The orbitals derived from the combination of $2p_y$ atomic orbitals are no longer symmetrical about the molecular axis. They are denoted $\pi_y 2p$ and π_y*2p. A similar pair of orbitals, $\pi_z 2p$ and π_z*2p, is formed by the combination of $2p_z$ atomic orbitals. Now wave functions which describe two or more orbitals of the same energy are called 'degenerate'; thus the $\pi 2p$ orbitals are doubly degenerate, since two orbitals of equal energy exist $(\pi_y 2p = \pi_z 2p)$; the $\pi*2p$ orbitals are also doubly degenerate $(\pi_y*2p = \pi_z*2p)$. Reference to *Figure 7.3* will show that, if a point in the orbital $\pi_y 2p$ is reflected in the centre, the sign of its co-ordinate changes, so ψ is written ψ_u. A similar reflection, in the centre, of points in the π_y*2p orbital shows that the sign of the co-ordinates is unchanged, and accordingly ψ is written ψ_g in this case. These distinctions are important because it can be shown that, when electrons change from one orbital to another, they only do so if they go into an orbital of different symmetry type. Transitions do not normally occur between two *gerade* orbitals or between two *ungerade* orbitals.

It is convenient to associate molecular orbitals with a new quantum number λ which is related to the m quantum number of atomic orbitals (page 37). The component of the angular momentum of the electron about the internuclear axis is $\lambda h/2\pi$, where λ takes the values $0, \pm 1, \pm 2, \ldots$ *etc*. The σ type molecular orbitals have $\lambda = 0$, and the π type orbitals have $\lambda = 1$. Combination of d atomic orbitals can give a δ molecular orbital for which $\lambda = 2$. These δ orbitals resemble 'double-π' orbitals in that they have two nodal planes intersecting in the molecular axis.

It might perhaps be wondered why we do not consider the combination of a $1s$ atomic orbital of one atom with the $2s$ atomic orbital of the other. However, in homonuclear molecules, where the atoms A and B are identical, the $1s$ and $2s$ atomic orbitals correspond to very different energies, whereas for effective combination the energies of the component wave functions must be comparable. For the same reason, s and p orbitals do not combine in homonuclear diatomic molecules. We shall see later in this chapter, however, that such combinations may be possible in heteronuclear molecules (*i.e.* when atoms A and B are different) since the energies of the orbitals may become comparable, although there are still certain limiting symmetry conditions.

Before we can discuss the electron arrangements (and apply the 'Aufbau' principle to molecules) in molecules other than the hydrogen molecule ion, we must arrange the molecular orbitals in order of their energies. These energies have been determined from spectroscopic observations, and when arranged in increasing order of magnitude, they give the sequence

$$\sigma 1s < \sigma^*1s < \sigma 2s < \sigma^*2s < \sigma 2p < \pi_y 2p = \pi_z 2p < \pi_y^*2p$$
$$= \pi_z^*2p < \sigma^*2p$$

A similar arrangement exists for the $\sigma 3s$ to σ^*3p molecular orbitals, but these energies are known with less certainty and, moreover, the bonding is now rarely between pure s, p, d . . . *etc.* atomic orbitals, but rather between certain combinations of them (*see e.g.* page 106). The terminology is a little cumbersome particularly for the π molecular orbitals, and Mulliken has suggested the following alternative form for the orbitals $\sigma 2s$ to σ^*2p

$$z\sigma < y\sigma < x\sigma < w\pi < v\pi < u\sigma$$

The notation can also be used for the 1-quantum (k shell) and 3-quantum (m shell) molecular orbitals by writing $(k)z\sigma$ and $(k)y\sigma$, and $(m)z\sigma$, $(m)y\sigma$, $(m)x\sigma$ . . . *etc.* Both notations have their own particular merits, and the student should be familiar with each of them. The Mulliken notation has two advantages in addition to its brevity, *viz.*

1. It may be applied to heteronuclear molecules, where the molecular orbital may be formed by combining two atomic orbitals from different quantum shells (*e.g.* HF, page 85, where the molecular orbital is compounded from $H(1s)$ and $F(2p_x)$ atomic orbitals).

2. When an orbital is written $\sigma 2s$, it is assumed that the electrons in it came originally from atomic $2s$ orbitals, but in the Mulliken notation a $z\sigma$ molecular orbital merely denotes the σ-type molecular orbital of lowest energy.

There are two important disadvantages of the terminology, however:

81

1. Possible confusion between z, y, and x and Cartesian co-ordinates (x, y, z) may arise.
2. The easily visualized relationship between the molecular orbital and the constituent atomic orbitals is lost.

We can now discuss the electronic distributions in simple homo-nuclear diatomic molecules by applying the 'Aufbau' principle. Electrons are assigned to the orbitals of lowest energy, the Pauli principle limiting each orbital to a maximum of two 'paired' electrons. (For molecular-orbital theory of polyatomic molecules *see* Chapter 9.)

HOMONUCLEAR DIATOMIC MOLECULES

1. H_2^+

We have already discussed the structure of the hydrogen molecule ion in detail, and have seen that the electron is in the bonding $\sigma 1s$ molecular orbital. The structure is accordingly written $H_2^+ [(\sigma 1s)]$.

2. H_2

The hydrogen molecule contains one electron more than the hydrogen molecule ion, and, in the ground state, this additional electron is also found in the $\sigma 1s$ orbital, giving the configuration $H_2[(\sigma 1s)]^2$. The hydrogen molecule thus contains two bonding electrons in the lowest energy bonding molecular orbital, $\sigma 1s$, and therefore, the two-electron bond of H_2 is much stronger than the one-electron bond of H_2^+.

3. He_2^+

This molecule has been detected spectroscopically. Its formation can be considered as the combination of a helium atom, He, and a helium ion, He^+, giving a configuration $(\sigma 1s)^2(\sigma^* 1s)$. The molecule differs from H_2 in that an extra electron is present in the antibonding $\sigma^* 1s$ orbital, and this antibonding electron reduces the bond strength: the 'relaxing' effect of an antibonding electron is rather greater than the binding force resulting from a bonding electron, so that the bond in He_2^+ is even weaker than a one-electron bond.

4. He_2

This molecule does not exist under normal conditions; its con-figuration would be $(\sigma 1s)^2(\sigma^* 1s)^2$, and the bonding power of

$(\sigma 1s)^2$ would be more than cancelled out by the antibonding $(\sigma^*1s)^2$. The non-existence of this molecule illustrates the general principle, *viz.* that the superposition of doubly filled bonding and antibonding orbitals leads to either no bonding or even to a slight overall repulsion.

Although the He_2 molecule is less stable than two uncombined helium atoms, it is, however, detected in discharge tubes by the spectrograph. Thus the combination of two excited helium atoms may lead to the formation of a He_2 molecule with the configuration $He_2[(\sigma 1s)^2(\sigma 2s)^2]$, *i.e.* with all four electrons in bonding orbitals. This configuration clearly corresponds to a higher energy than that of two normal uncombined helium atoms, but it appears to be more stable than two separate excited atoms.

5. Li_2

$$2Li(1s^2 2s) \rightarrow Li_2[KK(\sigma 2s)^2]$$

Here KK denotes the closed K shell structure $(\sigma 1s)^2(\sigma^*1s)^2$. The bonding between the two lithium atoms results from the pairing of the Li $2s$ electrons in the bonding $\sigma 2s$ orbital; the K shell electrons take virtually no part in the bonding, and remain essentially atomic in character. We shall generally assume that inner-shell electrons are non-bonding, although more refined treatments may have to consider inner-shell interactions.

The diatomic molecules of the other alkali metals have analogous configurations, $Na_2[KKLL(\sigma 3s)^2]$, . . . *etc.*; such diatomic molecules have been detected in the vapour state.

6. N_2

$$2N(1s^2 2s^2 2p^3) \rightarrow N_2[KK(\sigma 2s)^2(\sigma^*2s)^2(\sigma 2p)^2(\pi_y 2p = \pi_z 2p)^4]$$

or

$$N_2[KK(z\sigma)^2(y\sigma)^2(x\sigma)^2(w\pi)^4]$$

The K shell electrons again take no part in the bonding, and we have therefore two $2s$ and three $2p$ electrons from each nitrogen atom to feed into the available molecular orbitals (*cf. Figure 7.4*). The bonding resulting from $(z\sigma)^2$ is effectively cancelled out by the antibonding $(y\sigma)^2$, leaving the $(x\sigma)^2$ and $(w\pi)^4$ to provide the molecular bonding. These six bonding electrons produce a $N \equiv N$ triple bond, one bond being σ and the other two π in character. This triple bond is symmetrical about the N—N axis, the electronic charge cloud forming a cylinder around the bond.

The P_2 molecule has an analogous configuration, but with both

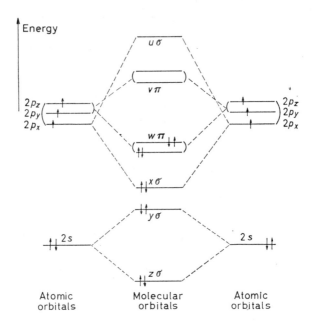

Figure 7.4. The formation of the molecular orbitals for N_2

the K and L inner shells full; the bonding electrons are the $[(m)x\sigma]^2$ and $[(m)w\pi]^4$, or $(\sigma 3p)^2$ and $(\pi_y 3p = \pi_z 3p)^4$.

7. O_2

$$2O(1s^2 2s^2 2p^4) \rightarrow O_2[KK(z\sigma)^2(y\sigma)^2(x\sigma)^2(w\pi)^4(v\pi)^2]$$

The oxygen molecule contains two electrons more than the nitrogen molecule, and these enter the antibonding $v\pi$ orbitals, so that the triple bond formed from $(x\sigma)^2$ and $(w\pi)^4$ is effectively reduced to a double bond by the antibonding character of $(v\pi)^2$. Since, however, there are two $v\pi$ antibonding molecular orbitals, then, by the rule that electrons go, if possible, into separate orbitals, with parallel spins (*see* Chapter 5, page 57), one electron will enter each of these, giving a paramagnetic molecule with two unpaired electrons. This explanation of the well known paramagnetism of the O_2 molecule, and an analogous explanation of the paramagnetism of S_2, was one of the early successes of the molecular-orbital theory.

84

8. F_2

$$2F(1s^2 2s^2 2p^5) \rightarrow F_2[KK(z\sigma)^2(y\sigma)^2(x\sigma)^2(w\pi)^4(v\pi)^4]$$

The $(z\sigma)^2$ and $(y\sigma)^2$ practically cancel one another out, as do the $(w\pi)^4$ and $(v\pi)^4$, so that although all fourteen electrons contribute in theory, the effective bonding is provided by the $(x\sigma)^2$. The F—F bond is therefore a single two-electron bond.

Cl_2 and Br_2 have structures analogous to F_2, Cl_2 having its inner K and L, and Br_2 its inner K, L, and M shells full.

9. Ne_2

$$2\,Ne(1s^2 2s^2 2p^6) \rightarrow Ne_2[KK(z\sigma)^2(y\sigma)^2(x\sigma)^2(w\pi)^4(v\pi)^4(u\sigma)^2]$$

This molecule is not formed under normal conditions because, as we saw in the corresponding example of the He_2 molecule, the number of bonding and antibonding electrons is equal; for similar reasons the other noble gas atoms do not form stable diatomic molecules.

HETERONUCLEAR DIATOMIC MOLECULES

The same general principles may be used in describing simple heteronuclear diatomic molecules, *i.e.* molecules in which the atoms A and B are no longer identical. We must take care in selecting the atomic orbitals that can combine effectively, and we recall that quantum theory requires three conditions to be satisfied for the effective combination of atomic orbitals. The atomic orbitals must *(1)* have similar energies, *(2)* have charge clouds that overlap as much as possible, and *(3)* have the same symmetry properties with respect to the axis A—B. We have already stressed the first condition in our discussion of homonuclear diatomic molecules, and since the atomic orbitals combined together were of the same type (*i.e.*, $1s$ with $1s$, $2s$ with $2s$, *etc.*), then the remaining two conditions were fulfilled automatically. The application of these principles to heteronuclear diatomic molecules can best be illustrated by a consideration of a number of simple examples.

1. HF *and the Other Hydrogen Halides*

The electronic configurations of hydrogen and fluorine atoms are

$$H \quad 1s$$

$$F \quad 1s^2\ 2s^2\ 2p^5$$

Now the molecular orbital describing the H—F bond must be compounded from a linear combination of the H($1s$) atomic orbital with one of the F atomic orbitals. Spectroscopic evidence shows that the energies of the $1s^2$ and $2s^2$ electrons in fluorine are far too low, however, to contribute appreciably to the bonding; this is merely another way of saying that the 'deep-seated' inner shell electrons do not take part in bonding, but remain essentially in atomic orbitals. The $2p$ electrons are the only ones of suitable energy; if we take the H—F axis as the x axis it can be shown that only the $2p_x$ orbitals give effective bonding. *Figures 7.5 (a)* and *7.5 (b)* show the effect of combining the hydrogen $1s$ atomic orbital with the fluorine $2p_x$ and $2p_y$ atomic orbitals, respectively (the overlap of the $2p_z$ orbital is exactly analogous with the $2p_y$).

The $2p_x$ orbital gives a much greater overlap with the hydrogen $1s$ orbital for a given internuclear distance than does the $2p_y$ orbital, and whereas in *(a)* the overlapping orbitals are of the same sign, in *(b)* the s orbital is overlapped by both lobes of the p orbital, so that the overlap from the positive lobe of the p orbital is exactly balanced by that of the negative lobe. These overlaps cancel one another out.

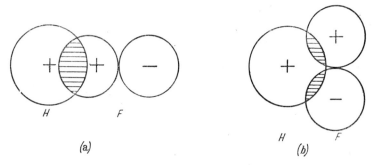

(a)   (b)

Figure 7.5. (*a*) Combination of the hydrogen $1s$ orbital with the fluorine $2p_x$ orbital, (*b*) combination of the hydrogen $1s$ orbital and the fluorine $2p_y$ orbital (overlap indicated by horizontal shading)

Thus the bonding between the hydrogen and fluorine atoms results almost entirely from the H($1s$) and F($2p_x$) electrons, and the other electrons in fluorine remain in atomic orbitals. We can write the configuration of HF as

$$\text{H}(1s) + \text{F}(1s^2 2s^2 2p^5) \rightarrow \text{HF}[\text{K}(2s)^2(x\sigma)^2(2p_y)^2(2p_z)^2]$$

and the bonding $x\sigma$ orbital is the only molecular orbital. This configuration closely resembles that of the neon atom $(1s^2 2s^2 2p^6)$ where all the orbitals are atomic. We must note carefully that, although the $x\sigma$ molecular orbital is compounded from the $H(1s)$ and $F(2px)$ atomic orbitals, the coefficients c_1 and c_2 in the molecular wave function $\psi = c_1 \phi_{H(1s)} + c_2 \phi_{F(2px)}$ are no longer equal. In this particular case $c_2 > c_1$, which implies that $\phi_{F(2px)}$ contributes more than $\phi_{H(1s)}$ to the molecular wave function ψ, thus giving a charge cloud with an electron density around the fluorine nucleus greater than that around the proton. The ratio c_2/c_1 (sometimes called λ) is a measure of the 'polarity' of the HF molecule. The other hydrogen halides have similar single-bonded structures, the $3p_x$, $4p_x$ and $5p_x$ orbitals being used by chlorine, bromine and iodine, respectively.

2. Molecules such as NO and CO

To a first approximation we can use the energy diagram of *Figure 7.4*. Carbon monoxide then has an identical configuration to the nitrogen molecule, and nitric oxide has one additional electron which is placed in the antibonding $v\pi$ orbital, *i.e.*

$$C(1s^2 2s^2 2p^2) + O(1s^2 2s^2 2p^4) \rightarrow CO[KK(z\sigma)^2 (y\sigma)^2 (x\sigma)^2 (w\pi)^4]$$
$$N(1s^2 2s^2 2p^3) + O(1s^2 2s^2 2p^4) \rightarrow NO[KK(z\sigma)^2 (y\sigma)^2 (x\sigma)^2 (w\pi)^4 (v\pi)]$$

Because of the single antibonding electron, the nitric oxide molecule is paramagnetic, and the bonding is weaker than in the nitrogen molecule. Nitric oxide is unexpectedly stable for a molecule with an unpaired electron, and this stability must be attributed to the electron being distributed over both atoms in a molecular orbital rather than localized on one atom.

The carbon monoxide configuration is exactly the same as that of the nitrogen molecule and implies a triple bond between the carbon and oxygen atoms. This description is not entirely adequate, however, because upon ionization, CO gives CO^+ and N_2 gives N_2^+, but whereas there is a bond weakening when N_2 is converted into N_2^+ (corresponding to a loss of one bonding electron), the bonding is stronger in CO^+ than in CO, as inferred by a shorter internuclear distance and a larger vibrational frequency. This is only understandable if CO initially contains a double bond which is changed towards a triple bond upon ionization. It has been suggested that one of the $w\pi$ orbitals may be largely atomic in character and centralized almost entirely over the oxygen atom.

We are not, of course, really justified in taking *Figure 7.4* as the energy diagram when the two bonded atoms have different electronegativities, because the a.o.'s* will not correspond to the same energy. Thus the energy of the 2s a.o. of oxygen will be appreciably less than that of the analogous carbon orbital. *Figure 7.6* indicates the sort of diagram that emerges, from which it may be seen that the bonding π orbitals now come below the bonding σ orbital. The carbon and oxygen a.o.'s no longer contribute equally

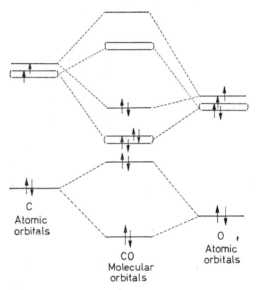

Figure 7.6. Molecular orbitals for carbon monoxide

to the m.o.'s; the coefficients of the oxygen a.o.'s are larger for the bonding m.o.'s and smaller for the antibonding m.o.'s. Thus the bonding electrons occupy molecular orbitals that are more 'concentrated' on the oxygen atom.

BOND STRENGTH

The examples so far discussed in this chapter show that stable molecules are formed when there is an excess of bonding electrons. The excess of bonding over antibonding electron pairs is called the 'bond order'. *Table 7.1* summarizes some evidence illustrating this point.

We can see that N_2 has a heat of dissociation roughly 3/2 times as great as that of O_2, while that of H_2 is roughly double that of

* We shall find it convenient henceforth to refer to atomic orbitals as a.o.'s and to molecular orbitals as m.o.'s.

Table 7.1. Relation of Excess of Bonding Electrons to Bonding Strength for Simple Diatomic Molecules[1]

Molecules	Pairs of bonding electrons	Pairs of anti-bonding electrons	Excess of bonding pairs	Bond dissociation energy (D_0) kcal/mole
CO	4	1	3	255·8
N_2	4	1	3	225·2
NO	4	$1\frac{1}{2}$	$2\frac{1}{2}$	149·9
O_2	4	2	2	117·97
H_2	1	0	1	109·5
H_2^+	$\frac{1}{2}$	0	$\frac{1}{2}$	64·4
He_2	1	1	0	—

H_2^+. The agreement between O_2 and H_2 is not good, but here it should be remembered, we are dealing with electrons of differing principal quantum number. Similarly, if we compare diatomic molecules involving second-row elements with their first-row analogues, we observe a decrease in bond dissociation energy. Thus the value for SiO is 184·8 kcal/mol. (*cf.* 255·8 for CO) and for S_2, 83·0 (*cf.* 117·97 for O_2). Evidently, electrons of higher principal quantum number have a smaller bonding power.

REFERENCE

[1] GLASSTONE, S. *Theoretical Chemistry*, p. 275. Van Nostrand, New York, 1945

For further details of the m.o. method see *e.g.*

GRAY, H. B. *Electrons and Chemical Bonding*, W. A. Benjamin Inc., New York and Amsterdam, 1964

8

THE VALENCE-BOND METHOD

IN the preceding chapter we discussed an approximation method for obtaining solutions to the appropriate Schrödinger equation for molecules. We considered the skeleton of the molecule (as given by the atomic nuclei, or nuclei plus inner shell electron) and evaluated the molecular energy levels that could be occupied by the electrons. These molecular levels or orbitals were obtained by taking a suitable linear combination of atomic orbitals. For clarity we restricted our discussion to diatomic molecules, but in Chapter 9 we shall see that, in principle, the ideas can be extended to polyatomic molecules, for which the electrons will be placed in molecular orbitals embracing all the atomic nuclei.

In this chapter we shall describe an alternative approximation procedure—the valence-bond method. Initially the discussion will be restricted to the simple diatomic molecules, the hydrogen molecule ion and the hydrogen molecule, but it will be extended to indicate how the method applies to more complicated molecules such as carbon dioxide and benzene. We shall see that the valence-bond method incorporates the idea of electron pairing, with each pair of electrons linking just two nuclei.

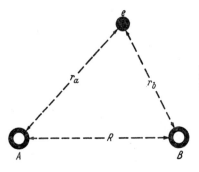

Figure 8.1. The hydrogen molecule ion: *A* and *B* represent two protons and *e* denotes the single electron

THE HYDROGEN MOLECULE ION

The Schrödinger equation for the molecule H_2^+ is

$$\nabla^2 \psi + \frac{8\pi^2 m}{h^2} \left(E + \frac{e^2}{r_a} + \frac{e^2}{r_b} - \frac{e^2}{R} \right) \psi = 0$$

This equation differs from that given for the hydrogen atom (page 33) in that the expression for the potential energy V now contains three terms, $- e^2/r_a$ and $- e^2/r_b$ which arise from the attraction of the electron for the nuclei A and B, and e^2/R which arises from the internuclear repulsion. The appropriate distances are shown in *Figure 8.1*.

When the two nuclei in this system are very far apart, we can consider two possible structures; in one structure the electron is associated entirely with the hydrogen nucleus A, nucleus B being a single proton (we denote this structure as $H_A H_B^+$ and describe it by a wave function ψ_I); in the other structure the electron is associated entirely with nucleus B and we assign a wave function ψ_{II} describing the structure $H_A^+ H_B$. These two structures correspond to states of equal energy, and if the nuclei were at infinite distance apart, they would give an accurate description of the electronic standing wave associated with either proton A or proton B. When the two nuclei are brought close together, however, neither structure adequately describes the situation, and we assume that a better description will be obtained by taking a linear combination of ψ_I and ψ_{II} so that the approximate wave function describing the molecule is

$$\psi = c_I \psi_I + c_{II} \psi_{II}$$

Since ψ_I and ψ_{II} represent states of equal energy, they contribute equally to ψ, and it can be shown that in this case there are two possible linear combinations

$$\psi_+ = \frac{1}{\sqrt{2}} \{\psi_I + \psi_{II}\}$$

and

$$\psi_- = \frac{1}{\sqrt{2}} \{\psi_I - \psi_{II}\}$$

The factor $1/\sqrt{2}$ is an approximate value of the normalizing constant, which ensures that $\int \psi^2 dv$ taken over the whole of space is unity. Now the charge density is proportional to ψ^2 and, therefore,

$$\psi_+^2 = \tfrac{1}{2}\{\psi_I^2 + \psi_{II}^2 + 2\psi_I \psi_{II}\}$$

while

$$\psi_-^2 = \tfrac{1}{2}\{\psi_I^2 + \psi_{II}^2 - 2\psi_I\psi_{II}\}$$

The ψ_+ function thus describes a state in which the charge density is greater than the sum of the separate charge densities, $\tfrac{1}{2}[\psi_I^2 + \psi_{II}^2]$, by an amount $\psi_I\psi_{II}$. These charge density distributions are shown in *Figure 8.2(a)* in which the broken lines represent the density functions ψ_I^2 and ψ_{II}^2, and the full lines ψ_+^2 and ψ_-^2; ψ_+^2 and ψ_-^2 can also be represented by contours—lines of equal electron density—as in *Figure 8.2(b)*. The diagrams for the ψ_+^2 function show clearly that the increased electron charge

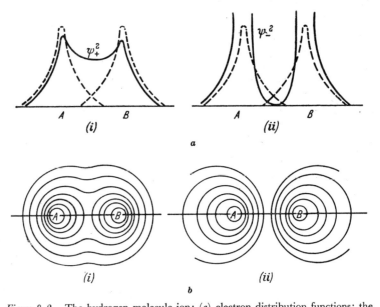

Figure 8.2. The hydrogen molecule ion: (*a*) electron distribution functions; the broken lines represent (i) ψ_I^2 or (ii) ψ_{II}^2, the full lines represent (i) ψ_+^2 or (ii) ψ_-^2
(*b*) electron contours, lines of equal electron density, for the states represented by (i) ψ_+^2 and (ii) ψ_-^2

density associated with this linear combination is concentrated in the region between the nuclei, whereas the charge cloud is pushed away from this region in the ψ_- combination.

Figure 8.3 shows how the energy varies with the internuclear distance. The function ψ_+ is evidently to be associated with the stable or 'ground' state of the H_2^+ molecule, since the curve shows an energy minimum at an internuclear separation or 'bond length' of r_0. The curve labelled ψ_- shows a steady increase of energy as the nuclei approach each other, no stable molecule is formed.

We can now summarize what happens when a proton approaches a hydrogen atom. When the nuclei are far apart, the electronic energy will be that of an electron in a single hydrogen atom, but as the nuclei get closer together, ordinary electrostatic (coulombic) forces will come into play—attraction between electron and protons, repulsion between proton and proton. Now an increase in the electron charge density in the region separating the protons will diminish the proton-proton repulsion, and, in general, the formation of a chemical bond can be associated with an increased electron charge density between the linked atoms.

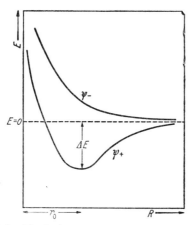

Figure 8.3. The hydrogen molecule ion: variation of energy with internuclear distance

In the case of the hydrogen molecule ion both the simple molecular-orbital and the valence-bond treatments lead to identical values for the energy and the bond length. Since ψ_I represents a structure in which the electron is associated only with nucleus A, it is an atomic orbital of A and identical with one of the component atomic orbitals (ϕ_A) of the LCAO treatment; similarly, ψ_{II} is identical with ϕ_B. Both methods use the linear combination technique, but the valence-bond treatment considers hypothetical 'structures', whereas the emphasis is on the electron in the molecular-orbital approach. The difference between the two methods will become more apparent when we consider more complicated molecules, for then the component wave functions of the linear combinations will no longer be the same.

THE HYDROGEN MOLECULE

The potential energy term V in the Schrödinger equation for the hydrogen molecule is

$$- e^2/r_{a_1} - e^2/r_{b_1} - e^2/r_{a_2} - e^2/r_{b_2} + e^2/R_{AB} + e^2/R_{12}$$

where the distances r_{a_1} etc., are those shown in *Figure 8.4*. The four negative terms correspond to the attraction of the electrons 1 and 2 for the protons A and B, and the two positive terms to repulsion between proton and proton, and between electron and electron.

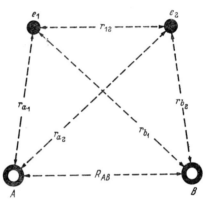

Figure 8.4. The hydrogen molecule: distances between electrons e_1 and e_2 and nuclei A and B

When the nuclei A and B are very far apart we have a structure $H_A{}^1 + H_B{}^2$, where $H_A{}^1$ represents electron 1 associated with nucleus A, and $H_B{}^2$ represents electron 2 associated with nucleus B. We can describe this system of two hydrogen atoms by a wave function ψ_I. An equally good description is afforded by a structure in which electron 2 is associated with nucleus A, and electron 1 with nucleus B, since the two electrons are indistinguishable, and this state, $H_A{}^2 + H_B{}^1$, is described by the wave function ψ_{II}. As the atoms are brought together, their standing waves will interact, and a new wave motion will result which Heitler and London suggested could be described as a linear combination of ψ_I and ψ_{II}, i.e.

$$\psi = \mathcal{N}\{c_I \psi_I + c_{II} \psi_{II}\}$$

Here again, ψ_I and ψ_{II} represent states of the same energy and they will be equally important in the linear combination; in fact $c_I = \pm c_{II}$, giving

$$\psi_+ = Nc_I\{\psi_I + \psi_{II}\}$$

and

$$\psi_- = Nc_I\{\psi_I - \psi_{II}\}.$$

A plot of energy against R_{AB} shows a minimum for the ψ_+ combination, corresponding to molecule formation. The calculated dissociation energy (3·14 eV) and bond length (0·87 Å) are roughly comparable with the experimentally determined values of 4·7466 eV and 0·74116 Å. The Heitler-London method can be improved by considering two additional structures in which both electrons are attached to one atom; these structures are the ionic forms

$$H_A^+ + H_B^- \text{ (or } H_A + H_B^{1, 2}) \text{ described by } \psi_{III}$$

and

$$H_A^- + H_B^+ \text{ (or } H_A^{1, 2} + H_B) \text{ described by } \psi_{IV}$$

where $H_A^{1, 2}$ represents nucleus A with two attached electrons. Thus Weinbaum used the trial function

$$\psi = \psi_I + \psi_{II} + c(\psi_{III} + \psi_{IV})$$

and applied the variation method to determine the value of the constant c that gave the state of lowest energy. These calculations gave a dissociation energy of 3·21 eV, and an equilibrium bond length of 0·884 Å for $c = 0·158$.

COMPARISON OF THE MOLECULAR-ORBITAL AND VALENCE-BOND METHODS

We can now usefully compare the molecular-orbital and the valence-bond descriptions of the hydrogen molecule, and to do this we must examine the component wave functions in more detail. Considering first of all the molecular-orbital method—if we neglect inter-electron repulsion, the probability of finding electron 1 in a volume dv_1 ($= \psi_1^2 dv_1$) is independent of the probability of finding electron 2 in a volume dv_2 ($= \psi_2^2 dv_2$), so that the probability that electron 1 is in the volume dv_1 while, simultaneously, electron 2 is in volume dv_2, is the product of the individual probabilities ($= \psi_1^2 dv_1 . \psi_2^2 dv_2$). The wave function ψ_+, which describes the molecule, is therefore given by the product $\psi_1 \psi_2$, i.e. the product of the functions for the individual electrons. Now we have seen (page 76) that in the molecular-orbital treatment, the wave function for a bonding electron in a molecule is given by the sum of atomic orbitals. Thus

$$\psi_1 = \phi_{A(1)} + \phi_{B(1)}$$

where $\phi_{A(1)}$ and $\phi_{B(1)}$ represent systems in which electron 1 is associated with nuclei A and B, respectively. (We are simplifying the expressions by omitting the normalizing constant and the coefficients c_A and c_B). Similarly, for electron 2,

$$\psi_2 = \phi_{A(2)} + \phi_{B(2)}$$

The overall molecular wave function is, therefore,

$$
\begin{aligned}
\psi_+ = \psi_1 \psi_2 &= (\phi_{A(1)} + \phi_{B(1)})(\phi_{A(2)} + \phi_{B(2)}) \\
&= \phi_{A(1)} \phi_{A(2)} + \phi_{B(1)} \phi_{B(2)} \\
&\quad + \phi_{A(1)} \phi_{B(2)} + \phi_{B(1)} \phi_{A(2)} \quad \ldots (43)
\end{aligned}
$$

Here the term $\phi_{A(1)} \phi_{A(2)}$ represents a structure in which both electrons are on nucleus A, thus producing ions H_A^- and H_B^+, while $\phi_{B(1)} \phi_{B(2)}$ represents the ionic form $H_A^+ H_B^-$.

In the valence-bond treatment, each constituent function of the linear combination represents a particular arrangement of protons and electrons, so that ψ_I, for example, refers to the structure in which electrons 1 and 2 are associated with nuclei A and B, respectively. Here again, the probability of finding electron 1 near nucleus A and, simultaneously, electron 2 on nucleus B, is given by the product of the individual probabilities, so that ψ_I can be written $\phi_{A(1)} \phi_{B(2)}$ and ψ_{II} can be written $\phi_{A(2)} \phi_{B(1)}$. Hence, taking the linear combination,

$$\psi_+ = \phi_{A(1)} \phi_{B(2)} + \phi_{A(2)} \phi_{B(1)} \quad \ldots (44)$$

Equation 44 differs from equation 43 in the absence of the 'ionic' terms $\phi_{A(1)} \phi_{A(2)}$ and $\phi_{B(1)} \phi_{B(2)}$. The simple (Heitler-London) valence-bond theory assumes, in effect, that these structures will not be important, because the mutual repulsion between electrons would reduce the probability that they would be found simultaneously close together on the same nucleus. The molecular-orbital theory, on the other hand, neglects the effect of interelectronic repulsion, and gives equal weight to the ionic and non-ionic (*i.e.* covalent) terms. It is clear that neither method will give reliable results unless it is suitably modified, because each describes limiting conditions, and it can indeed be shown that as the two methods are improved they 'converge' and finally become equivalent. The modifications involve the consideration of more elaborate wave functions, with correspondingly more elaborate calculations. The recent treatment of the hydrogen molecule by KOLOS and ROOTHAAN (1960) uses a function containing 50 terms, and calculations of this kind have only become feasible because of the development of high-speed electronic computers. *Table 8.1* shows values for the inter-nuclear distance and the binding energy for H_2^+ and H_2 as calculated by a number of workers, and as determined experimentally. These values show that the two-electron bond in H_2 is much stronger and shorter than the one-electron bond in H_2^+.

Table 8.1. Bond Lengths and Binding Energies for H_2^+ and H_2

Method	H_2^+		H_2	
	D (eV)	R (Å)	D (eV)	R (Å)
Experimental	2·791	1·060	4·7466	0·74116
Simple LCAO	1·76	1·32	2·681	0·850
James (1935)	2·772	—		
Heitler-London (1927)			3·140	0·869
Weinbaum (1933)			3·21	0·884
Kolos and Roothaan (1960)			4·7467	0·74127

These results also show that complete accuracy is only obtained by abandoning the simple LCAO functions, but, as Coulson has pointed out, this also involves abandoning conventional chemical concepts and simple pictorial quality in the results. The great advantage of the admittedly inaccurate linear combination method is that it can be related to the chemist's familiar pictures of bond diagrams and structural formulae.

RESONANCE

So far we have tried to avoid using the word 'resonance' which has long been associated with the valence-bond theory. We have described the hydrogen molecule by a wave function which is a linear combination of two functions ψ_I and ψ_{II}, each of which describes a hypothetical state or structure, *i.e.* $H_A^1 H_B^2$ and $H_A^2 H_B^1$. The actual state of the molecule is often described as being a resonance hybrid of the states represented by ψ_I and ψ_{II}

or, more simply, that there is resonance between the structures $H_A{}^1 H_B{}^2$ and $H_A{}^2 H_B{}^1$. The danger of using this terminology is that it is almost impossible to avoid giving the impression that the hypothetical structures have a real molecular existence. In the case quoted, the structures $H_A{}^1 H_B{}^2$ and $H_A{}^2 H_B{}^1$ correspond to separated atoms, and do therefore have a real existence, but we must not imagine that there is some kind of rapid exchange of electron between the nuclei; in other words there is no kind of equilibrium between the structures.

In the more complicated molecules that we shall be discussing shortly, the contributing structures will not, in general, have any real existence. They are, in fact, merely a convenient device for getting an approximate solution of a complicated wave equation, but they also have the great practical advantage in that they enable the chemist to picture molecule formation in terms of conventional and familiar bond diagrams, even though the actual state of the molecule cannot be described directly by a single conventional structural formula.

The selection of hypothetical structures from which the true structure is compounded by taking a linear combination is not an arbitrary matter. If possible structures have wave functions ψ_I, ψ_{II} ... etc., the important ones will be those for which c is large in the combination $\psi = N\{c_I\psi_I + c_{II}\psi_{II} + \dots\}$ and a detailed study of the expression for the energy associated with the function ψ shows that effective contributions are made by structures that

(a) have similar energies,

(b) have, approximately, the same relative positions of the nuclei,

(c) have the same numbers of unpaired electrons.

These conditions, are, in fact, developed from those already mentioned on page 76 where we are discussing linear combinations in terms of the molecular-orbital theory. We saw there that effective combination only occurs when the component functions represent states of almost equal energy, when their orbitals overlap to a considerable extent, and when the orbitals have the same symmetry with respect to a molecular axis.

These points can be illustrated by discussing a specific example— the nitromethane molecule—for which two alternative bond diagrams can be drawn if we regard the nuclei as occupying fixed positions in space—

$$CH_3\!-\!\overset{+}{N}\!\!\underset{O^-}{\overset{\diagup O}{\diagdown}}\qquad \text{Structure } I$$

$$CH_3\!-\!\overset{+}{N}\!\!\underset{O}{\overset{\diagup O^-}{\diagdown}}\qquad \text{Structure } II$$

From a study of a range of nitrogen–oxygen molecules it has been established that a nitrogen–oxygen double bond is much shorter than a single bond (*see* Chapter 11), but experimental work on nitromethane has shown that both nitrogen–oxygen bonds are of the same length (1·22Å). The actual state of the molecule is described in terms of hypothetical structures I and II, and we write the wave function for the molecule as the linear combination

$$\psi = \mathcal{N}\{c_I\psi_I + c_{II}\psi_{II}\}$$

We can picture this as a superposing and blending of the structures I and II. Each structure contributes equally to the combination, since the energies of each are the same, they have the same relative positions of the nuclei, and the same number (zero) of unpaired electrons. In Chapter 9 we shall see how molecular-orbital theory can be used to give a more adequate picture of bonding in this molecule.

Our second example will be carbon dioxide, which has long been given the structural formula $O{=}C{=}O$. This seems to satisfy conventional valency requirements, but the carbon–oxygen bond length is found to be 1·15Å, whereas the length of a normal carbon-oxygen double bond (as in ketones) is 1·22Å, and that calculated for triple bond is 1·10Å. The bond in carbon dioxide thus seems to be intermediate in character between a double and triple bond, and this is explained by describing the structure in terms of three hypothetical forms

$$O{=}C{=}O \quad \ldots \quad \psi_I$$
$$\overset{+}{O}{\equiv}C{-}\overset{-}{O} \quad \ldots \quad \psi_{II}$$
$$\overset{-}{O}{-}C{\equiv}\overset{+}{O} \quad \ldots \quad \psi_{III}$$

so that $\psi = \mathcal{N}\{c_I\psi_I + c_{II}\psi_{II} + c_{III}\psi_{III}\}$ describes the actual state of the molecule.

The inadequacy of the $O{=}C{=}O$ formulation is also demonstrated by energy considerations. The energy of a carbon–oxygen double bond is known to be 175 kcal, so that the heat of formation of the molecule $O=C=O$ should be approximately twice this value, *i.e.* 350 kcal; in fact, the measured heat of formation is 383 kcal, and the difference between these two values, 33 kcal, represents the extent to which the actual structure is more stable than that represented by the simple formula $O=C=O$. This difference in energy between the actual molecule and the most stable of the hypothetical structures is called the resonance energy (*see Figure 8.5*).

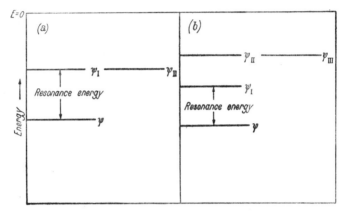

Figure 8.5. Resonance energy: (*a*) for hydrogen, (*b*) for carbon dioxide

The benzene molecule affords perhaps the best example of the way in which the resonance theory can explain some apparently very curious facts—namely, that benzene is extremely stable chemically, although the conventional structural formula introduced by Kekulé includes three double bonds, that the molecule is completely planar, and that the carbon–carbon bonds all have the same length. We can write down five conventional structural formulae for the molecule

each of which can be represented as before by an appropriate wave function ψ_I, ψ_{II}, ... *etc.* The actual structure of the benzene molecule is then described by a wave function ψ given by

$$\psi = c_I \psi_I + c_{II} \psi_{II} + c_{III} \psi_{III} + c_{IV} \psi_{IV} + c_V \psi_V$$

in which the contributions of the two Kekulé forms will be more important than those of the Dewar forms. (The Dewar structures will be less stable, *i.e.* of higher energy, because of the weak long bonds across the ring.) The bonds between adjacent carbon atoms in the resulting resonance hybrid are now all equivalent, being neither single nor double but being 'flavoured' with the characteristics of each. In fact, the measured carbon–carbon distance in benzene is 1·39Å, whereas the single bond length measured in the saturated hydrocarbons is 1·54 Å, and the double bond length in ethylene is 1·32 Å.

RESONANCE—SOME MISCONCEPTIONS AND SOME GUIDING PRINCIPLES

It is important to stress that the structures which are written down to describe molecules, and which are superposed in the valence-bond treatment, have no real existence. Benzene does not contain equal proportions of molecules with structures

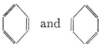

nor is there a tautomeric equilibrium

set up, in which electrons oscillate between one position and another, nor can it be said that benzene exists for a certain fraction of time in the form

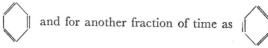

The molecule itself has a single structure which cannot be described directly by a conventional bond diagram. A molecule which is a resonance hybrid has been compared with a mule; the mule has characteristics of the donkey and the horse, but at no time is it either of these.

Resonance theory has been in danger of falling into disrepute because of the uncritical way in which the concept has been applied. It is only too easy to explain away difficulties in interpreting the behaviour of molecules by inventing a large number of improbable structural formulae and saying that the molecule is a hybrid constructed from these forms. The hypothetical structures that are used to form the linear combination must always obey the rules formulated on page 98. Thus the rule that the structures must represent the same relative position of the nuclei illustrates the fundamental difference between resonance and tautomerism. A molecule such as acetone, for instance, exists in the so-called keto and enol forms

$$CH_3—C—CH_3 \qquad\qquad CH_3—C=CH_2$$
$$\underset{\text{keto}}{\overset{\|}{O}} \qquad\qquad\qquad \underset{\text{enol}}{\overset{|}{OH}}$$

which differ not only in the allocation of the bonding electrons but also in the position of the hydrogen atom—bonded to the carbon in the keto form and to the oxygen in the enol form. There is an actual equilibrium between the two tautomeric forms, which are definite entities and can usually (always in theory) be separated from one another. The amount of any tautomer present in the equilibrium depends on its energy content; thus the enol form of acetone has a much higher energy than the keto form, and is present in only very small amounts. The fact that tautomers actually exist while the resonance forms do not, can be further illustrated by a consideration of dipole moments. Nitrous oxide may be considered as a hybrid of the two forms $\overset{-}{N}=\overset{+}{N}=O$ and $\overset{+}{N}\equiv N-\overset{-}{O}$, and since the dipoles of the contributing structures have opposed orientation, the resultant dipole of the actual molecule is very small. If, however, these two forms were tautomers (differing only in electron displacement in this case), then both structures would align themselves in the applied field so as to give a high mean dipole moment.

The rule that the contributing structures must be of comparable energy can be exemplified by benzene, where the Dewar forms have a much higher energy, and are therefore less important, than the more stable Kekulé forms in the combination that represents the actual molecule. In the same way, the simple Heitler-London treatment of the hydrogen molecule can be modified to include contributions from ionic structures such as $H_A^{1\,2} + H_B$, and $H_A + H_B^{1\,2}$, but these, being of higher energy, contribute less than the forms $H_A^1 + H_B^2$ and $H_A^2 + H_B^1$ to the actual structure.

102

9

DIRECTED VALENCY

THE CRITERION OF MAXIMUM OVERLAPPING

IN our discussion of the hydrogen molecule by the valence-bond and molecular-orbital methods, we saw that bond formation was accompanied by the overlapping of atomic orbitals. This concept is of great importance in discussing directed valency and we must now consider it more fully. Both the approximation methods give an expression for the energy of the stable bond which involves a definite integral of the form $\int \phi_A \phi_B dv$, where the integration is taken over the whole of space. When ϕ_A and ϕ_B are small, the integral will be very small; it only becomes significant in regions where ϕ_A and ϕ_B have appreciable values at the same instant— that is when the atomic orbitals occupy the same region of space.

For this reason $\int \phi_A \phi_B dv$ is called the overlap integral. We also saw in Chapter 7, however, that overlapping of atomic orbitals only results in stable bond formation if certain other conditions are satisfied, *viz.* that the atomic orbitals are of similar energy and appropriate symmetry. The latter limitations arise because certain orbitals (*e.g. p, d*) have lobes of different sign separated by nodal planes. An example of this is discussed on page 86 of Chapter 7. These symmetry limitations must always be borne in mind when the overlap of orbitals is being considered.

Now we can get a very useful qualitative explanation of directed valency if we assume that the strength of a covalent bond is approximately proportional to the amount of overlap. Let us consider some overlapping orbitals from this point of view.

1. Overlap of *s*-type orbitals

The orbitals are spherically symmetrical so that they overlap to the same extent in all directions.

2. Overlap of *p*-type orbitals

We recall from our treatment on page 85 of Chapter 7 that the bond in the F_2 molecule can be ascribed to the overlap of the $2p_x$ atomic orbitals.

103

In this case, the electron density is greatest along the x axis, and so we get maximum overlap for a given internuclear distance if the bond is formed in this direction. Since a large overlap is associated with the formation of a strong bond, we may perhaps relate bond strength to the type of orbital from which the bond is formed. It was pointed out by Pauling that, since the p orbitals are concentrated along particular directions, they should overlap more effectively for a given internuclear distance than should s orbitals of the same principal quantum number. If the radius of a $3s$ orbital is r, the length of the axis of a $3p$ orbital, from nucleus to boundary, can be shown to be $\sqrt{3} \cdot r$; a bond formed by the overlap of a p orbital should be approximately $\sqrt{3}$ times as strong as one formed by the corresponding s orbital. Too much significance must not be attached to the numerical factor, but the general principle remains a useful one.

These ideas can now be used to discuss the structure of poly-atomic molecules.

The oxygen atom, for instance, has the configuration

$$(1s)^2(2s)^2(2p_x)^2(2p_y)(2p_z)$$

with its two unpaired electrons in orbitals which are at right angles to one another. Let us picture the approach of two hydrogen atoms to the oxygen atom to form a molecule of water. As one hydrogen atom approaches (say along the z axis), the $2p_z$ orbital becomes distorted and finally overlaps the s orbital of the hydrogen atom. The s orbital of the second hydrogen atom overlaps the oxygen $2p_y$ orbital in the same way; it must moreover approach at an angle of 90° to the first OH bond if an appreciable overlap is to result. In other words, the oxygen atom forms two bonds at right angles to one another, since greatest overlap—and hence the strongest bonds—are then obtained. In practice, we find that the HOH angle is not the predicted 90°, but 104·5°; this difference has been ascribed to the mutual repulsion of the hydrogen atoms. When we consider the related molecules H_2S and H_2Se, however, we find the bond angles are 93° and 91°, respectively, i.e. almost the predicted behaviour, since the bond lengths are greater and the mutual repulsion of the hydrogen atoms correspondingly diminished.

In a similar way, we get three mutually perpendicular bonds in the NH_3, PH_3, AsH_3, and SbH_3 molecules as a result of the over-lap of the s orbitals of the respective hydrogen atoms with the three p orbitals of nitrogen, phosphorus, arsenic, and antimony. Here again, the bond angles are not exactly 90°, being 107·3° for NH_3,

93·3° for PH_3, 91·8° for AsH_3, and 91·3° for SbH_3; as the bonds get longer the repulsions between the hydrogen atoms diminish and the bond angle approaches 90°. An alternative description of the structures of these molecules is given on page 114.

POLYATOMIC MOLECULES—V.B. AND M.O. DESCRIPTIONS

If we try to apply the ideas of the previous section to a discussion of bonding in compounds of elements such as beryllium, boron and carbon, we at once run into difficulties. Beryllium, with a ground state $(1s)^2(2s)^2$, might be expected to behave as a noble gas such as argon and form no bonds at all. If, however, the beryllium atom receives sufficient energy, one of its $2s$ electrons may be promoted into a $2p$ orbital, giving the excited state $(1s)^2(2s)(2p_x)$, with two unpaired electrons. These could then pair with electrons from two atoms (X), giving rise to a BeX_2 molecule. Normally, we should expect the energy of the two Be—X bonds to be greater than the promotion energy. We can use either the valence-bond or molecular-orbital methods to describe such molecules, and these are now discussed in some detail, first for linear triatomic molecules (e.g. BeX_2), then for both trigonal-planar and tetrahedral molecules.

1. Linear Triatomic Molecules

(a) Valence-bond approach—If we take the specific case of a $BeCl_2$ molecule* then at a very simple level we can visualize the bonding in two stages, firstly by the $2p_x$ orbital and then by the $2s$ orbital (Figure 9.1). The beryllium $2p_x$ orbital overlaps with a chlorine $3p_x$ orbital to form a strong well-defined σ bond, the overlap being best when the p orbitals of the beryllium and chlorine atoms are collinear. The second Be—Cl bond is less clearly defined, however, because the Be $2s$ orbital is spherically symmetrical and accordingly overlaps equally well in any direction. Mutual repulsion between the two chlorine atoms and between the two bonding pairs of electrons will undoubtedly tend to give a large Cl–Be–Cl angle, but no precise value can be predicted on grounds of maximum overlap. Our naive picture of the formation of two Be—Cl bonds from dissimilar beryllium orbitals thus describes a structure with ill-defined bond angles and bonds of unequal strength.

It is well established, however, that both bonds in such simple compounds of bivalent beryllium are collinear. It seems necessary,

---

* We assume the molecule is monomeric in solution or in the gas phase; solid $BeCl_2$ is a polymeric substance (see page 159).

therefore, to describe this equality of bonding in terms of two entirely equivalent orbitals used by the beryllium atom. Thus we suppose that the beryllium atom does not use simple $2s$ and $2p_x$ orbitals but a combination of these. When we describe the electronic states of the beryllium atom, we do so, in theory, by solving the appropriate wave equation and obtaining suitable solutions ψ_{2s} and ψ_{2p_x}, each of which describes an energy state or orbital capable of accommodating two electrons with opposed spins. Now valid

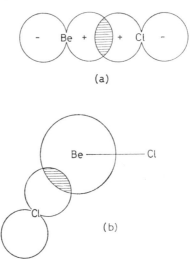

(a)

(b)

Figure 9.1. The hypothetical bonding in monomeric $BeCl_2$ in terms of overlap of (*a*) p orbital of beryllium with p orbital of chlorine, (*b*) Be s orbital with Cl p orbital

solutions to the wave equation are also given by linear combinations of these wave functions, and we can describe other possible orbitals by combined or 'hybridized' wave functions. If the two solutions, ψ_{2s} and ψ_{2p_x}, have equal weight, we get two new equivalent functions which describe two linearly-directed orbitals:

$$\psi_{sp(i)} = \sqrt{\tfrac{1}{2}}(\psi_{2s} + \psi_{2p_x})$$
$$\psi_{sp(ii)} = \sqrt{\tfrac{1}{2}}(\psi_{2s} - \psi_{2p_x})$$

In these combinations, similar to those described in Chapter 8, $\sqrt{\tfrac{1}{2}}$ is the normalization constant.

In wave-mechanical language we say that two sp hybrid orbitals have formed from one s and one p orbital. These new hybrid orbitals (sometimes called 'equivalent' orbitals) have strong directional characteristics, and each protrudes further along the axis than the original contributing p orbital.

Figure 9.2 shows how these two hybrids, which point in diametrically opposite directions, arise from the combinations shown in the above equations. Thus $\psi_{sp(i)}$ has a large lobe with $+$ sign on the right because the $+$ signs of the s and p orbitals coincide here;

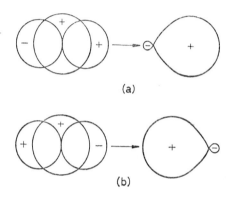

(a)

(b)

Figure 9.2. The formation of (a) $\psi_{sp(i)}$, (b) $\psi_{sp(ii)}$

for $\psi_{sp(ii)}$, the combination involves $-\psi_{2p_x}$, which inverts the signs on the $2p_x$ orbital lobes, so that the large lobe with the $+$ sign is now on the left. Since the hybrid orbitals protrude further than the s and p orbitals, we should expect them to provide more effective overlapping and stronger bonds.

In a colloquial way, we can describe the hybridization by saying that the electron probability clouds of the pure s and p orbitals have interacted to produce new standing waves by the merging and reforming of the charge clouds. The charge cloud of the p orbital is concentrated along an axis, and the merger with the s orbital charge cloud extends this axial concentration. The important point is that the two electrons 'occupy' a certain region of space and that we can describe this space either as two sp hybrid orbitals or as an s and a p orbital; the total charge cloud remains symmetrical about the axis.

The formation of the $BeCl_2$ molecule can now be visualized as resulting from the overlap of the sp hybrid orbitals of beryllium with the $3p_x$ orbitals of the chlorine atoms. For the most effective overlap, the two Be—Cl bonds must now be collinear and equivalent (*cf.*

Figure 9.3). This description of the arrangement of the two bonding electron pairs about the beryllium atom in terms of hybrid orbitals is another example of the working of the exclusion principle, which requires electrons of similar spin to be as far away from each other as possible (see page 53).

At this point we should perhaps make some reservations about colloquial phrases of the type 'two entirely equivalent orbitals *used* by the beryllium atom . . .' which have already appeared in this chapter. There is the danger of ascribing real physical significance to the orbitals, as though they were electron 'containers' hooked on

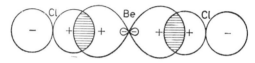

Figure 9.3. The overlap of the $3p_x$ orbitals of Cl with sp hybrid orbitals of Be in the molecule $BeCl_2$

to the atom. One should say, rather, that the charge-cloud distribution around the beryllium atom in a molecule such as $BeCl_2$ is such that it can be described mathematically in terms of collinear, equivalent hybrid orbitals. Having made this reservation we shall, however, for the sake of brevity continue to use the colloquial form of words.

(*b*) *Molecular-orbital approach*—In the valence-bond description we combined the wave functions describing the $2s$ and $2p_x$ atomic orbitals to give functions describing two sp hybridized atomic orbitals. Overlapping of these hybrid orbitals with the $3p_x$ orbitals from the chlorine atoms gives two localized two-electron bonds; each bond links beryllium with just one chlorine atom.

An alternative way of describing the bonding is to use the concepts of molecular orbitals that were developed in Chapter 7. We recall that in the description of the hydrogen molecule the combination of two a.o.'s (one from each atom) yielded two m.o.'s, one bonding and one antibonding. In general, the combination of n a.o.'s (n even) yields n m.o.'s, $n/2$ bonding and $n/2$ antibonding. Thus in a linear triatomic molecule ($BeCl_2$), the beryllium atom has two suitable a.o.'s ($2s$ and $2p_x$) and each chlorine atom has one ($3p_x$); the combination of these four a.o.'s will accordingly give four m.o.'s, two bonding and two antibonding. The combinations, which are expressed mathematically below, are shown pictorially in *Figure 9.4* and energetically in *Figure 9.5*.

$$\text{Cl}_a\text{–Be–Cl}_b$$

(i) $\quad \sigma_s = \psi_{Be(2s)} + \psi_{Cl_a} + \psi_{Cl_b}$

(ii) $\quad \sigma_s = \psi_{Be(2s)} - \psi_{Cl_a} - \psi_{Cl_b}$

(iii) $\quad \sigma_p = \psi_{Be(2p_x)} - \psi_{Cl_a} + \psi_{Cl_b}$

(iv) $\quad \sigma^* = \psi_{Be(2p_x)} + \psi_{Cl_a} - \psi_{Cl_b}$

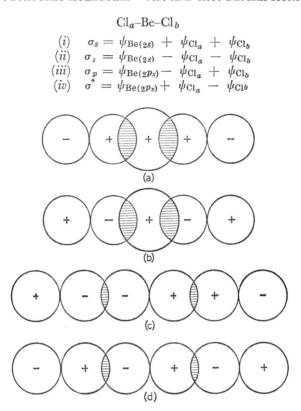

Figure 9.4. The formation of molecular orbitals for $BeCl_2$

In the pictorial diagram we use the convention that the p orbital of beryllium has a $+$ sign for its right-hand lobe; the chlorine p orbitals have their lobes with a $+$ sign pointing towards the beryllium. In the energy diagram we have assumed for simplicity that the a.o.'s of the chlorine atoms will be of similar energy to those of the $2s$ and $2p_x$ a.o.'s of beryllium, and this is reflected in the wave functions (i)–(iv) where the a.o.'s are mixed equally. Of course, in the general case of MX_2 molecules the energies of the M and X orbitals may be somewhat different, and the molecular-orbital wave functions will then incorporate appropriate mixing coefficients, just as they did for hydrogen fluoride (see page 86). There will, however, always be two bonding m.o.'s, each delocalized and embracing the three nuclei, and each containing two electrons. Thus,

109

on the molecular-orbital approach we describe the bonding in a linear triatomic molecule as consisting of two delocalized two-electron bonds.

If we compare the valence-bond and molecular-orbital approaches for such molecules, we see that the electron density pattern is the same in each case and has a symmetrical distribution about the x axis. We can use either method to describe this density pattern, since both tell us that the three atoms are collinear, bonded by four electrons.

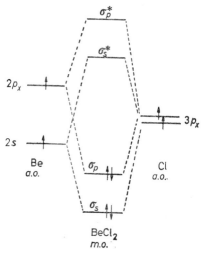

Figure 9.5. Simplified molecular-orbital energy levels for $BeCl_2$

The $2p_y$ and $2p_z$ atomic orbitals of Be have been omitted, since they are of the wrong symmetry to contribute to the bonding: *see* page 79.

2. *Trigonal-planar Molecules*

In molecules such as boron trichloride we have to account for the experimental observation that the three bonds are coplanar with bond angles of $120°$. If we consider the boron atom, with configuration $(1s)^2(2s)^2(2p_x)$, we see that promotion of one $2s$ electron gives the excited state $(1s)^2(2s)(2p_x)(2p_y)$ with three unpaired electrons; these three unpaired electrons give rise to the three bonds, by pairing with an unpaired electron from each of the three chlorine atoms. As with the linear molecules (*e.g.* $BeCl_2$), there are two convenient ways of discussing this bonding.

(*a*) *Valence-bond approach*—The $2s$, $2p_x$, and $2p_y$ orbitals are mixed or hybridized to give three equivalent sp^2 orbitals. The mathematical treatment shows that the state of lowest energy corresponds to three orbitals which are coplanar and oriented at 120° to each other:

$$\psi_{sp^2(i)} = \sqrt{\tfrac{1}{3}}\psi_{2s} + \sqrt{\tfrac{2}{3}}\phi_{2p_x}$$
$$\psi_{sp^2(ii)} = \sqrt{\tfrac{1}{3}}\psi_{2s} - \sqrt{\tfrac{1}{6}}\phi_{2p_x} + \sqrt{\tfrac{1}{2}}\psi_{2p_y}$$
$$\psi_{sp^2(iii)} = \sqrt{\tfrac{1}{3}}\psi_{2s} - \sqrt{\tfrac{1}{6}}\phi_{2p_x} - \sqrt{\tfrac{1}{2}}\psi_{2p_y}$$

It should be noted that ψ_{2p_y} makes no contribution to $\psi_{sp^2(i)}$ because we have chosen the x axis as axis for this hybrid orbital. The

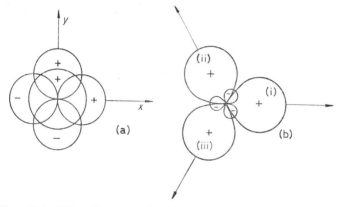

Figure 9.6. Trigonal-planar molecules: (*a*) unhybridized 2s, $2p_x$ and $2p_y$ orbitals; (*b*) hybridized sp^2 orbitals

coefficients differ for the contribution of ψ_{2p_x} and ψ_{2p_y} to the other two hybrids, since the angles made by the x and y axes to the hybrid orbital axes are 60° and 30°, respectively. *Figure 9.6* illustrates the formation of the hybrid orbitals from the component orbitals. The process can be visualized as the merging and re-forming of the s and p charge clouds to give the sp^2 hybrid charge clouds. This concentration of the charge clouds in a plane is as might be expected, since the original $2p_x$ and $2p_y$ orbitals were co-planar. The boron–chlorine bonds are σ bonds and can be described as the overlap of the sp^2 hybrid orbitals of the boron atom with the $3p_x$ atomic orbitals of the chlorine atoms.

(*b*) *Molecular-orbital approach*—We use the same method as for the linear triatomic molecules, except that we now have four atoms and six a.o.'s (2s, $2p_x$ and $2p_y$ for B, and $3p_x$ for each of the three Cl atoms), with accordingly six m.o.'s, three bonding (σ_s, σ_x, σ_y) and three antibonding (σ_s^*, σ_x^*, σ_y^*). The energy diagram is shown in

Figure 9.7, where for simplicity we assume that the energy of the chlorine $3p$ orbitals is between that of the $2s$ and $2p$ orbitals of boron. *Figure 9.8* shows how the σ_s m.o. is obtained by the combination of the boron $2s$ orbital with the $3p$ orbital of each of the three chlorine atoms. It also shows the formation of one of the σ_p m.o.'s from the boron $2p_x$ orbital; here it can be seen that overlap of the $2p_x$ orbital is most effective with the p orbital of chlorine atom 1, and this would

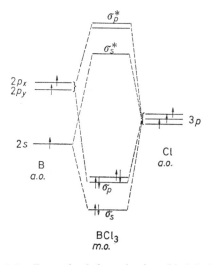

Figure 9.7. Energy levels for molecular orbitals in BCl_3

be reflected by giving a greater weighting to this chlorine a.o. in the molecular-orbital function. Thus, in the function

$$\psi\sigma_{p_x} = \psi_{B2p_x} + c_1\psi_{Cl_1} - c_2(\psi_{Cl_2} + \psi_{Cl_3})$$

c_1 is greater than c_2. There are three bonding m.o.'s, each embracing four nuclei and containing two electrons. Hence in this molecular-orbital description we consider the bonding to consist of three two-electron four-centred bonds. In both of these approaches the total bonding involves six electrons, and the electron-density charge-cloud patterns would in either case have a planar symmetry.

3. *Tetrahedral Molecules*

We may use methane as a typical example of a molecule with a tetrahedral configuration. As with beryllium and boron, we acquire the necessary number of unpaired electrons by promoting a $2s$ electron, giving carbon the configuration

$$(1s)^2(2s)(2p_x)(2p_y)(2p_z),$$

in which the electron charge cloud is spherically symmetrical. The four C—H bonds can be described by the v.b. or m.o. technique.

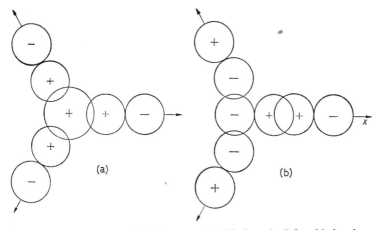

Figure 9.8.   Formation of (a) the σs m.o. in BCl_3 from the B $2s$ orbital and a $3p$ orbital from each Cl atom; (b) one of the σp m.o.'s from the B $2p_x$ orbital and the Cl p orbitals

(a) *Valence-bond approach*—The concept of four localized two-electron bonds is readily realized by describing the four unpaired electrons on the carbon atom in terms of four sp^3 hybrid orbitals which point towards the corners of a tetrahedron. This is expressed mathematically by taking linear combinations of the $2s$ and $2p$ orbitals:

$$\psi_{sp^3}(i) = \tfrac{1}{2}(\psi_{2s} + \psi_{2p_x} + \psi_{2p_y} + \psi_{2p_z})$$
$$\psi_{sp^3}(ii) = \tfrac{1}{2}(\psi_{2s} + \psi_{2p_x} - \psi_{2p_y} - \psi_{2p_z})$$
$$\psi_{sp^3}(iii) = \tfrac{1}{2}(\psi_{2s} - \psi_{2p_x} + \psi_{2p_y} - \psi_{2p_z})$$
$$\psi_{sp^3}(iv) = \tfrac{1}{2}(\psi_{2s} - \psi_{2p_x} - \psi_{2p_y} + \psi_{2p_z})$$

We then describe the methane molecule in terms of the overlap of the s orbitals of the hydrogen atoms with the sp^3 hybrid orbitals of the carbon atom. Inasmuch as the hybrid orbitals are largely concentrated along the tetrahedral axes, the C—H bonds will also be formed along these axes in order to get the maximum overlap for a given internuclear distance.

Ethane is described in much the same way, with each carbon atom using four sp^3 hybrid orbitals; the C—C bond is described as the overlap of an sp^3 hybrid orbital from each carbon atom and the C—H bonds as the overlap of the carbon sp^3 and hydrogen s orbitals.

The structures of many simple compounds of nitrogen and oxygen can also be discussed in terms of hybrid orbitals, and this treatment has some advantages over that described on pages 104, 105. We can, for instance, regard the five outer electrons of nitrogen $(2s^22p^3)$ as occupying four approximately tetrahedral hybrid orbitals, three orbitals being singly occupied and the other doubly occupied (giving a so-called 'lone pair'). Overlap of the singly-filled orbitals with hydrogen $1s$ orbitals gives the ammonia molecule, in which the bond angles are approximately tetrahedral, *i.e.* 109·5°. Thus there are three two-electron localized bonds and one lone-pair of (non-bonding) electrons.

Similarly, the six outer electrons of oxygen $(2s^22p^4)$ can be described by four tetrahedral hybrid orbitals, two being singly occupied and two containing a pair of electrons, and the water molecule formed by the overlap of hydrogen $1s$ orbitals with the singly-occupied oxygen orbitals should then yield a bond angle close to the predicted tetrahedral value. We shall discuss these ideas in considerably more detail in Chapter 11.

(*b*) *Molecular-orbital approach*—We use the same general principles already applied to linear and trigonal-planar molecules, except that we now have to consider four a.o.'s for the central atom, one s and three p, together with an a.o. from each of the four univalent atoms. Hence the combination of eight a.o.'s gives eight m.o.'s four bonding and four antibonding. *Figure 9.9* shows a typical energy level diagram scheme for CH_4, where once again we have assumed that all the a.o.'s are of similar energy. The eight electrons, four from carbon and one from each hydrogen, are placed in the four bonding m.o.'s. If, for simplicity, we take all the weighting coefficients as unity—that is, regard all the a.o.'s as being of the same energy— then the linear combinations are the following:

$$\sigma_s = \psi_{C_{2s}} + (\psi_{Ha_{1s}} + \psi_{Hb_{1s}} + \psi_{Hc_{1s}} + \psi_{Hd_{1s}})$$
$$\sigma_s^* = \psi_{C_{2s}} - (\psi_{Ha_{1s}} + \psi_{Hb_{1s}} + \psi_{Hc_{1s}} + \psi_{Hd_{1s}})$$
$$\sigma_{px} = \psi_{C_{2px}} + (\psi_{Ha_{1s}} + \psi_{Hb_{1s}} - \psi_{Hc_{1s}} - \psi_{Hd_{1s}})$$
$$\sigma_{px}^* = \psi_{C_{2px}} - (\psi_{Ha_{1s}} + \psi_{Hb_{1s}} - \psi_{Hc_{1s}} - \psi_{Hd_{1s}})$$
$$\sigma_{py} = \psi_{C_{2py}} + (\psi_{Ha_{1s}} - \psi_{Hb_{1s}} + \psi_{Hc_{1s}} - \psi_{Hd_{1s}})$$
$$\sigma_{py}^* = \psi_{C_{2py}} - (\psi_{Ha_{1s}} - \psi_{Hb_{1s}} + \psi_{Hc_{1s}} - \psi_{Hd_{1s}})$$
$$\sigma_{pz} = \psi_{C_{2pz}} + (\psi_{Ha_{1s}} - \psi_{Hb_{1s}} - \psi_{Hc_{1s}} + \psi_{Hd_{1s}})$$
$$\sigma_{pz}^* = \psi_{C_{2pz}} - (\psi_{Ha_{1s}} - \psi_{Hb_{1s}} - \psi_{Hc_{1s}} + \psi_{Hd_{1s}})$$

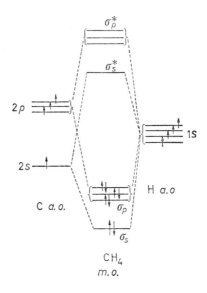

Figure 9.9. Energy level diagram for m.o.'s in CH_4

4. *Molecules using d Orbitals for Bonding*

Elements in the second row of the Periodic Table and beyond, where the electrons begin to occupy d levels, may form compounds showing symmetries other than linear, trigonal-planar or tetrahedral. By far the commonest configuration is the octahedral one, found in such compounds as SF_6 and $[Co(NH_3)_6]^{3+}$. Five co-ordination, which is observed in simple compounds such as PF_5 and in co-ordination compounds such as $VCl_3,2NMe_3$, leads to two basic arrangements, trigonal bipyramidal or square pyramidal; there are several alternative stereochemical arrangements for seven and eight co-ordination, and these will be discussed in some detail in Chapters 11 and 12. The latter chapter will also give an account of the square-planar arrangement that is sometimes found in preference to the tetrahedral one when the central quadrivalent atom forming the four bonds is a transition element.

Once again it is possible to use either a valence-bond or molecular-orbital approach to describe the bonds in any of these configurations. The concept of hybridization, with localized two-electron σ bonds, is the simplest picture, but the more sophisticated molecular-orbital approach is generally necessary for the consideration of properties such as spectra and magnetic susceptibility. In Chapter 12 we shall see that there is a third description, the ligand field one, which is related to the molecular-orbital description but is somewhat simpler; it is applicable to transition metal compounds where the d electrons play a vital role.

For the moment, however, we are more concerned with the simple stereochemistry of compounds, and for this purpose the valence-bond approach is adequate. The octahedral arrangement is described in terms of sp^3d^2 hybrid orbitals, obtained by taking linear combinations of the wave functions describing the s, p_x, p_y, p_z, $d_{x^2-y^2}$ and d_{z^2} orbitals:

$$\psi_{(i)} = \sqrt{\tfrac{1}{6}}\ \psi_s + \sqrt{\tfrac{1}{2}}\ \psi_{p_z} + \sqrt{\tfrac{1}{3}}\ \psi_{d_{z^2}}$$
$$\psi_{(ii)} = \sqrt{\tfrac{1}{6}}\ \psi_s - \sqrt{\tfrac{1}{2}}\ \psi_{p_z} + \sqrt{\tfrac{1}{3}}\ \psi_{d_{z^2}}$$
$$\psi_{(iii)} = \sqrt{\tfrac{1}{6}}\ \psi_s + \sqrt{\tfrac{1}{12}}\ \psi_{d_{z^2}} + \tfrac{1}{2}\psi_{d_{x^2-y^2}} + \sqrt{\tfrac{1}{2}}\ \psi_{p_x}$$
$$\psi_{(iv)} = \sqrt{\tfrac{1}{6}}\ \psi_s + \sqrt{\tfrac{1}{12}}\ \psi_{d_{z^2}} + \tfrac{1}{2}\psi_{d_{x^2-y^2}} - \sqrt{\tfrac{1}{2}}\ \psi_{p_x}$$
$$\psi_{(v)} = \sqrt{\tfrac{1}{6}}\ \psi_s + \sqrt{\tfrac{1}{12}}\ \psi_{d_{z^2}} - \tfrac{1}{2}\psi_{d_{x^2-y^2}} + \sqrt{\tfrac{1}{2}}\ \psi_{p_y}$$
$$\psi_{(vi)} = \sqrt{\tfrac{1}{6}}\ \psi_s + \sqrt{\tfrac{1}{12}}\ \psi_{d_{z^2}} - \tfrac{1}{2}\psi_{d_{x^2-y^2}} - \sqrt{\tfrac{1}{2}}\ \psi_{p_y}$$

It should be noted that the six hybrid orbitals point along the x, y and z axes, as might be expected since the atomic orbitals used in their construction also did this; thus we used the d_{z^2} and $d_{x^2-y^2}$ orbitals rather than the d_{xy}, d_{xz} and d_{yz}. *Figure 9.10* shows the combination producing $\psi_{(i)}$; $\psi_{(ii)}$ merely involves changing signs on the p_z orbital and gives a hybrid orbital of the same shape but

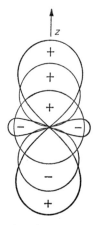

Figure 9.10. Combination of s, p_z and d_{z^2} orbitals to give $\psi_{(i)}$

116

pointing in the opposite direction. The reader can see for himself how the functions $\psi_{(iii)} - \psi_{(vi)}$ give orbitals pointing along the x and y axes. The d orbitals used in the hybridization can originate from the penultimate shell *i.e.* $(n-1)$ d^2, ns, np^3, as in $[Co(NH_3)_6]^{3+}$, or from the same shell as the s and p orbitals, *i.e.* ns, np^3, nd^2, as in SF_6; the hybrid orbitals have the same shape in either case. To avoid confusion in later discussion involving the use of these orbitals in complex formation, we will designate them d^2sp^3 and sp^3d^2, respectively.

Table 9.1 gives a summary of the commoner hybrid orbitals together with typical examples of molecules in which they occur. They will be discussed fully in Chapters 11 and 12.

Table 9.1. Hybrid Orbitals

	Simple			Hybridized		
Type	Degree of protrusion	Examples	Type	Degree of protrusion	Shape	Examples
s	$1 \cdot 0$	H_2	sp	$1 \cdot 93$	linear	$HgCl_2$
p	$1 \cdot 73$	Cl_2	sp^2	$1 \cdot 99$	trigonal-planar	BCl_3
d	not used alone		sp^3	$2 \cdot 00$	tetrahedral	CH_4
			dsp^2	$2 \cdot 69$	square-planar	K_2PtCl_4
			d^2sp^3	$2 \cdot 93$	octahedral	Rb_2TiCl_6
			dsp^3	$2 \cdot 80$	trigonal-bipyramidal	PCl_5
			d^4sp^3	—	dodecahedral	$K_4Mo(CN)_8$

So far, when we have referred to molecules having certain shapes such as trigonal-planar or tetrahedral, the examples we have chosen have been those in which the multivalent atom is bonded with a number of identical atoms. Thus boron trichloride is trigonal-planar with Cl–B–Cl angles of 120°, and methane is tetrahedral with all the H–C–H angles the same. When the bonded atoms are not the same, however, the bond angles are slightly different, *e.g.*

$$CH_4 \text{ all H–C–H angles } = 109 \cdot 5°$$
$$CCl_4 \text{ all Cl–C–Cl angles } = 109 \cdot 5°$$
$$CH_3Cl \text{ all H–C–H angles } = 110 \cdot 9°$$
$$CHCl_3 \text{ all Cl–C–Cl angles } = 110 \cdot 9°$$

and although these variations are small, they are nevertheless significant. In these cases some of the hybrid orbitals have a little more p character than others, *i.e.* they are not all exactly equivalent. Many more examples of this type of variation from an 'ideal' bond angle will be met with in Chapter 11.

So far in this chapter we have restricted our discussion of molecular shape to molecules containing σ bonds only, and we have pointed out that while the valence-bond description involves the appropriate number of localized two-electron bonds, the molecular-orbital approach describes the behaviour of electrons by means of orbitals that are delocalized. We saw in Chapter 7 that when more than two electrons hold two nuclei together, then the orbitals may be divided into σ and π type depending upon their symmetry. In the N_2 molecule, for instance, bonding involves six electrons in one σ and two π orbitals (*cf.* page 83).

If we now return to a consideration of the simple compounds of carbon, we see that in a discussion of the structure of ethylene, both σ- and π-type bonds appear. Thus it is observed experimentally that all six atoms are co-planar with bond angles close to 120°, and it is convenient to consider that carbon makes use of two of its three $2p$ orbitals and the $2s$ orbital to form three co-planar sp^2 hybrid bonds; the remaining $2p_z$ orbital, which is not hybridized, has its axis perpendicular to the plane (xy) of the hybrid orbitals, one lobe being above the plane, the other below. A σ-type carbon–carbon bond is formed by the overlap of two sp^2 hybrid orbitals, one from each atom, and the four carbon–hydrogen bonds are formed by the overlap of the remaining sp^2 hybrid orbitals with hydrogen s orbitals; all the nuclei are therefore co-planar. There is also the possibility of *lateral* overlap of the two unhybridized p_z orbitals, and while this is not as great as the end-on type of overlap, it will be appreciable because the axes of these p orbitals are parallel. The second bond between the carbon atoms is thus a π bond. It is weaker than the σ bond and its presence accounts for the reactivity of ethylene and of other compounds containing similar localized π bonds. Any attempt to twist the molecule about the C–C axis would result in a decrease in the overlap of the $2p_z$ orbitals and would in effect, break the π bond. The absence of free rotation of the methylene groups about the C=C bond is thus explained.

It is important to realize that this double bond has two regions of charge density, one above and one below the plane of the molecule, and that these two charge clouds belong to the same orbital and cannot be considered independently, *i.e.* the plane of the molecule is the nodal plane of the π orbital.

With the acetylene molecule (*cf. Figure 9.11*), the valence-bond description takes carbon *sp* hybrid orbitals as the basis for the

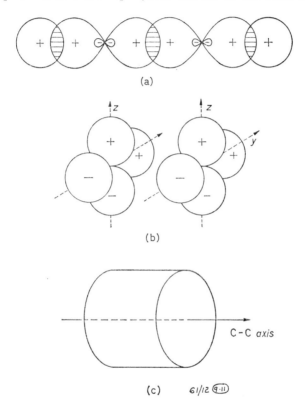

Figure 9.11. Orbital overlap in acetylene: (*a*) formation of σ bonds; (*b*) p_y and p_z orbitals which overlap to give π bonding; (*c*) charge cloud distribution in the π bonds

formation of the H–C–C–H skeleton, leaving two of the $2p$ orbitals ($2p_y$ and $2p_z$) unchanged. Lateral overlap of these p orbitals gives two π bonds between the carbon atoms, the total charge distribution (for the four electrons) having cylindrical symmetry about the C–C axis [*cf. Figure 9.11(c)*].

The π bonds we have just mentioned for ethylene and acetylene are localized between two carbon atoms, but there are many molecules in which the π orbitals may be delocalized and embrace more than two nuclei. Indeed, it is just such molecules that are so difficult to represent on paper by conventional structural formulae. Amongst many examples may be mentioned benzene, where we had to resort

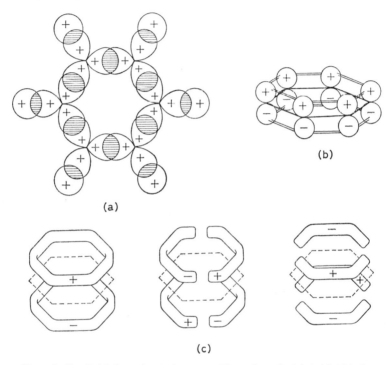

Figure 9.12. Orbital overlap in benzene: (*a*) overlap of sp^2 hybrid orbitals forming σ bonds; (*b*) overlap of $2p_z$ orbitals; (*c*) the delocalized bonding π orbitals

to the use of at least two structural forms to emphasize that all the bonds were equal. We know from experiment that benzene is a regular hexagon with bond angles of 120°, so we may use a valence-bond description for the framework σ bonds [*cf. Figure 9.12(a)*] in which carbon is sp^2 hybridized, as in ethylene. We now have to consider the six p orbitals perpendicular to the plane of the carbon hexagon. Lateral overlap of these p orbitals, taken in pairs, will give three localized π bonds, corresponding to one or other of the Kekulé structures, but there is no reason why the overlap should be limited in this way. Thus the $2p_z$ orbital on any carbon atom in

Figure 9.12(b) can overlap equally well with the $2p_z$ orbital on either of the neighbouring carbon atoms. The π bonding is therefore considered to result from mutual overlapping of all the $2p_z$ orbitals, giving delocalized molecular orbitals [*Figure 9.12(a)*] associated with all the carbon nuclei in the molecule. Because there are six atomic orbitals there will be six molecular orbitals, three bonding and three antibonding, and the bonding orbitals will accom-

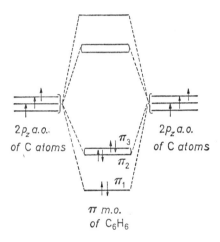

Figure 9.13. Energy level diagram for m.o.'s in benzene

modate the six π electrons. *Figure 9.13* shows the energy scheme for the π electrons in benzene. The lowest energy bonding π orbital provides a charge cloud consisting of two hexagonal shaped 'streamers', one above and one below the plane of the benzene ring. The other two bonding π orbitals, which are of higher energy, have an additional node and may be described colloquially as 'split hexagonal streamers'.

Since the π electrons are now in delocalized π orbitals, they are of lower energy than they would be in the localized orbitals implied by Kekulé structures showing alternate single and double bonds. The energy difference is the delocalization energy, which corresponds to the resonance energy of the valence-bond theory.

We saw in Chapter 8 that there were many molecules such as benzene in which conventional formulae did not adequately express the bonding. With nitromethane, for instance, we had to use two structural formulae to indicate that both N—O bonds were equivalent. In this molecule the bonds attached to the nitrogen atom are co-planar, so that in valence-bond terminology we can regard the σ bonding as arising from nitrogen sp^2 hybrids. As we shall see in the next chapter, the basic shapes of molecules are determined by their σ bonds and, therefore, may be described simply in terms of hybridization. Any delocalized π bonding may then be expressed

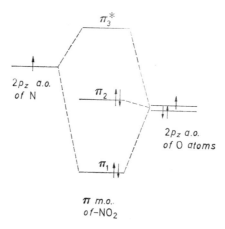

Figure 9.14. Energy level diagram for π bonding in nitromethane

in molecular-orbital terms. Thus in nitromethane the nitrogen atom and the two oxygen atoms have their $2p_z$ orbitals perpendicular to the plane of the nitro group, so that the combination of the three a.o.'s will give rise to three m.o.'s. If to a first approximation we consider the nitrogen and oxygen $2p_z$ a.o.'s to have the same energy, then we should expect one bonding, one non-bonding and one antibonding m.o. *Figure 9.14* shows the energy diagram for this π bonding, corresponding to the following three wave functions

$$\psi_{(i)} = \psi_{N(\pi)} + \psi_{O1(\pi)} + \psi_{O2(\pi)}$$
$$\psi_{(ii)} = \psi_{O1(\pi)} + \psi_{O2(\pi)}$$
$$\psi_{(iii)} = \psi_{N(\pi)} - \psi_{O1(\pi)} - \psi_{O2(\pi)}$$

The four π electrons will occupy $\psi_{(i)}$ and $\psi_{(ii)}$, so that the essential bonding is a three-centre two-electron π bond. The two non-bonding electrons will not contribute to the stability of the molecule; they correspond to the resonance description of the electrons being in the $2p_z$ a.o. of either oxygen atom.

Figure 9.15 shows diagrammatically how the $\psi_{(i)}$ m.o. arises.

It is, of course, perfectly possible to put down the full m.o. diagram for the $-NO_2$ grouping, which is merely the superposition of the σ energy diagram for a trigonal-planar molecule (*cf. Figure 9.7*) and the π diagram of *Figure 9.14*. In principle, we should always consider this full diagram, but we shall find (see next Chapter) that it is more convenient in many cases to consider the π bonding separately on an m.o. diagram and use the hybridization approach for σ bonds.

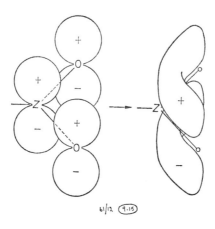

61/12 (9.15)

Figure 9.15. Formation of the $\psi_{(i)}$ m.o. from the p_z orbitals of nitrogen and oxygen

THE EQUIVALENT-ORBITAL DESCRIPTION

Up till now we have described multiple bonds, as in ethylene and acetylene, in terms of σ and π orbitals, but it is perfectly possible to discuss these structures in terms of hybrid orbitals that are, approximately, tetrahedrally arranged (see *e.g.* POPLE[1]). In ethylene, the carbon–hydrogen bonds are formed by the overlap of carbon sp^3 hybrid orbitals with hydrogen $1s$ orbitals, and the double bond results from the overlap of the remaining two hybrid orbitals on each carbon atom; this latter overlap (*cf. Figure 9.16*) gives two 'bent' bonds or 'banana' orbitals disposed above and below the plane containing the carbon and hydrogen nuclei, each orbital containing two electrons. The formation of a double bond involves some distortion of the original hybrid orbital distribution, and the angle between the two C—H bonds in ethylene is somewhat more than the tetrahedral value of $109 \cdot 5°$. Although this description

123

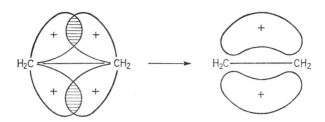

Figure 9.16. Ethylene: formation of the double bond from overlap of carbon sp^3 hybrid orbitals

may appear to be wildly different from that of σ and π bonds, it is merely another way of 'dividing up' the four bonding electrons into two regions. This is done mathematically by rearranging the wave functions. Thus the σ and π bond description involves two functions

$$\psi_\sigma = \frac{1}{\sqrt{2}}\,(\psi c_{1(\sigma)} + \psi c_{2(\sigma)})$$

$$\psi_\pi = \frac{1}{\sqrt{2}}\,(\psi c_{1(\pi)} + \psi c_{2(\pi)})$$

where the σ a.o.'s are 'sp' hybrids and the π a.o.'s are the $2p_z$ orbitals. If we take linear combinations of ψ_σ and ψ_π we get two equivalent orbitals

$$\psi_{(i)} = \frac{1}{\sqrt{2}}\,(\psi_\sigma + \psi_\pi)$$

$$\psi_{(ii)} = \frac{1}{\sqrt{2}}\,(\psi_\sigma - \psi_\pi)$$

This is illustrated diagrammatically in *Figure 9.17*.

We shall refer back to this approach in Chapter 13 when discussing the bonding in diborane.

The triple bond between the carbon atoms in acetylene can also be represented by 'banana' or 'bent' bonds, in this case three, each containing two electrons. The reader will, no doubt, realize that this representation of two-carbon atom linking by one, two or three bonds is the orbital presentation of the old idea of tetrahedra linking through a point, a side or a face.

BOND ENERGIES

At the beginning of this chapter we related the strength of a bond to the extent of the overlap of the orbitals involved in its formation.

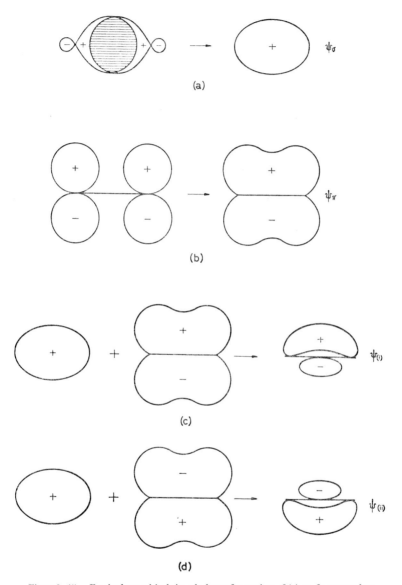

Figure 9.17. Equivalent orbitals in ethylene: formation of (a) ψ_σ from overlap of C σ orbitals (*sp* hybrids); (b) ψ_π from overlap of C p_z orbitals; (c) $\psi_{(i)}$ from $\psi_\sigma + \psi_\pi$; (d) $\psi_{(ii)}$ from $\psi_\sigma - \psi_\pi$

Unfortunately this idea, which gives an easily visualized description of bond formation, breaks down when we consider the sp type of hybrid orbital. Thus if we give the radius of an s orbital the arbitrary value unity, the axes of the hybrid orbitals, measured from the nucleus, have the following lengths: $sp = 1\cdot93$, $sp^2 = 1\cdot99$ and $sp^3 = 2$. We should, therefore, expect the bond strengths to increase slightly in this order, but WALSH[2] has shown that the order is, in fact, the reverse of this. Table 9.2 (a) gives values for the C—H bond length and C—H bond energy in suitable molecules.

Table 9.2 (a)

Molecule	Hybridiza-tion	C–H bond length (Å)	C–H bond energy kcal/mole
Acetylene	sp	$1\cdot057$	~121
Ethylene	sp^2	$1\cdot079$	~106
Methane	sp^3	$1\cdot094$	~103
CH radical	(p)	$1\cdot120$	80

The shortening of the C—H bond as the 's character' of the hybrid orbital increases is paralleled in other bonds, and BROWN[3] has correlated bond length and hybrid character for C—X bonds, where X = carbon, halogens, oxygen and nitrogen. The way in which C—C bond lengths vary with changes in hybridization is shown in Table 9.2 (b).

Table 9.2 (b)

Molecule	Hybridization	C—C bond length (Å)
Ethane	sp^3—sp^3	$1\cdot54$
Propene	sp^3—sp^2	$1\cdot51$
Methyl acetylene	sp^3—sp	$1\cdot46$

Coulson has pointed out that while Pauling's comparison of bond length and overlap is very useful in a qualitative sense, it is not adequate for the quantitative consideration of bond energies in the sp^3, sp^2, sp hybrids, and it would in fact suggest zero energy for π bonds. A better estimate of the bond strength is given in terms of the overlap integral. The concept of maximum overlap does, however, remain a most useful one in establishing the directions in which bonds are formed, and hybrid orbitals of the type

d^2sp^3, dsp^2, *etc.*, which do protrude extensively along their axes, do certainly form very stable bonds.

REFERENCES

[1] POPLE, J. A. *Q. Rev. chem. Soc.* 11 (1957) 273
[2] WALSH, A. D. *Trans. Faraday Soc.* 43 (1947) 60
[3] BROWN, M. G. *Trans. Faraday Soc.* 55 (1959) 694

10

IONIC, METALLIC AND HYDROGEN BONDS

IT may eventually be possible to describe all molecules by a single comprehensive theory of chemical bonding, but, at the present time, it is more convenient to classify bonds into a number of types—covalent, ionic, metallic, molecular and hydrogen bonds. We have dealt with the theory of the covalent bond in Chapters 7, 8 and 9, and we now discuss the other bond types.

IONIC BONDS

The Kossel, Lewis, and Langmuir theories explained electrovalency, and the formation of electrovalent (*i.e.* ionic) bonds in terms of the electrostatic attraction between ions of opposite charge, these ions being formed by a complete transfer of electrons between atoms. The formation of ionic bonds is thus related to the ease with which ions can be formed from neutral atoms, and to the way in which the ions are packed together in the crystal structure.

The formation of positive ions from neutral atoms has already been discussed (page 55) in terms of ionization energy. *Figure 5.4* shows that the alkali metals will be the most likely elements to lose one electron each and form stable cations, but even in these cases the amount of energy needed is quite large—*ca.* 120 kcal/mole. The formation of negatively charged ions is usually discussed in terms of 'electron affinity', the energy liberated when a neutral atom acquires an electron and forms a stable negative ion. Electron affinity can also be regarded as the ionization energy of a negative ion, *i.e.* the energy needed to remove one electron from the singly-charged ion

$$Cl^- + \text{'electron affinity'} \rightarrow Cl + 1e^-$$

Electron affinity is not easy to measure experimentally. Some values for halogens and also for oxygen, $O \rightarrow O^-$, are quoted in *Table 10.1*. They are taken from a useful review article by PRITCHARD and SKINNER[1].

A decrease in value with increasing ionic radius would be ex-

Table 10.1.   Electron Affinities

Atom	F	Cl	Br	I	O
Electron affinity, kcal/mole	83·7	87·2	81·7	74·7	33·5

pected in a series such as the halide ions, but there is as yet no adequate explanation of the anomalous low value for fluorine.

Energy is liberated when the halogen atoms form halide ions with single negative charges. These ions will have the stable noble gas configuration, but the singly charged O^- ion would have to acquire a second electron to achieve this stable configuration. However, in the formation of the stable oxide ion, O^{2-}, more energy is absorbed in overcoming the repulsion between the second electron and the O^- ion than is supplied by the $O \rightarrow O^-$ conversion; there is, therefore, a net absorption of energy (167·9 kcal/mole) in the change from oxygen atom to oxide ion. We observe that the numerical values of electron affinities are, in general, smaller than those of the ionization potentials; thus the energy released in the formation of a negative ion may not be sufficient to remove an electron from another atom to form a positive ion. In the formation of an electrovalent compound, $A^+ B^-$, from two atoms A and B, only part of the ionization energy of A is supplied by the electron affinity of B; however, more than sufficient energy is supplied by the electrostatic attraction between the ions when they are brought close together. This electrostatic attraction may be very large in ionic crystals, which consist of a regular arrangement of positive and negative ions in a crystal lattice. (Crystals are built up from small units packed together side by side; the points defining the corners of these units produce what is known as a simple crystal lattice.) The energy required to remove the ions from their equilibrium position in the crystal to infinity is called 'crystal' or 'lattice' energy. Methods of calculating crystal (lattice) energies are not discussed in this book; *see e.g.* the review article by WADDINGTON[2].

In recent years (*see* review articles by BAYLISS[3] and by WALSH[4]), measurements of ionization potentials of molecules have been used to obtain the relative strengths of different types of bond. Thus, the first ionization potential of ethylene, which corresponds to the removal of an electron from a π orbital, is 242·2 kcal/mole, whereas the first ionization potential of acetylene is 263·2 kcal/mole. The π bonds in acetylene are, therefore, stronger than those in ethylene. Such measurements are of particular value for those organic molecules in which the carbon atoms use three hybrid sp^2 orbitals and one p orbital (*cf.* benzene, page 120). We assume that two of the six electrons of such a carbon atom remain close to the nucleus in the K shell (*i.e.* they have the $1s^2$ configuration); three of the four remaining electrons are used in the formation of σ bonds, and the fourth goes into a delocalized π orbital. Measurements of ionization potentials show us that the removal of a K shell electron requires an energy of about 6,459 kcal/mole, removal of a σ electron about 323 kcal/mole, and removal of a π electron 230·7 kcal/mole. This gives us the experimental justification for neglecting the behaviour of the K shell and σ electrons in discussing molecular structures of this type. We concentrate our attention on the π electrons since, being much less firmly attached to the carbon nuclei, these are largely responsible for the reactivity of the molecules.

THE SIZE OF IONS

The size of ions is determined by the attractive force exerted on the outer electrons by the effective nuclear (positive) charge*. Thus, when a neutral atom is converted into a positive ion, we should expect the size to decrease, since there has been a net increase in the effective nuclear charge. Conversely, we should expect a negative ion to be larger than the neutral atom from which it has been formed.

Table 10.2. Interionic Distances

Interionic distance (Å)	KF 2·66 NaF 2·31	KCl 3·14 NaCl 2·81	KBr 3·29 NaBr 2·98	KI 3·53 NaI 3·23
Δ(KF–NaF *etc.*)	0·35	0·33	0·31	0·30

The distance between neighbouring positive and negative ions in crystals can be measured by x-ray diffraction methods; details of the method are outside the scope of the present work; (*see e.g.* WHEATLEY[5], LONSDALE[6]). Some values of the interionic distances in a series of alkali metal halides are given in *Table 10.2*. If we subtract the interionic distance in a sodium halide from the interionic distance in the corresponding potassium compound, we get an almost constant value for the difference, Δ, as shown in the table.

* The effective nuclear charge is the true nuclear charge diminished by the effect of the inner or 'screening' electrons.

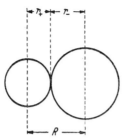

Figure 10.1. Interionic distance and ionic radii

This is most simply explained if we assume that each ion acts as a sphere of constant radius; the measured interionic distance R *(Figure 10.1)* is then the sum of the radii of two spheres in contact, $r_+ + r_-$. It may seem surprising that this assumption works so well, for we emphasized in earlier chapters that, although the electron charge cloud is concentrated in shells close to the nucleus, there is still a finite probability of finding an electron at a considerable distance from the nucleus. However, reference to *Figure 5.1*, page 50, which illustrates the radial distribution of charge density in the sodium ion, shows that this density decreases very rapidly as r increases, so that there is little error in assuming that the ion has a finite radius. We can, therefore, use experimentally determined values of interionic distances to obtain the radii of all other ions if the radius of one ion is already known. Thus, from *Figure 10.1*, if R and r_+ are known, r_- is given by:

$$r_- = R - r_+$$

The values of ionic radii now in general use are based on a semi-empirical method devised by PAULING[7a] (1927). He divides the distance between a pair of isoelectronic ions, such as K^+ and Cl^-, in the inverse ratio of their effective nuclear charges. The screening effect of the inner electrons can be calculated (approximately); thus, for K^+ and Cl^-, the screening constant is $10.87e$, where e is the electronic charge, so that the effective nuclear charges for these ions are $(19 - 10.87) e = 8.13 e$ and $(17 - 10.87) e = 6.13 e$, respectively. The interionic distance in KCl is 3.14 Å, and dividing this distance in the inverse ratio of $8.13 : 6.13$ we get the radii of the K^+ and Cl^- ions as follows:

$$K^+ = \frac{6.13}{6.13 + 8.13} \times 3.14 = 1.35 \text{ Å}$$

$$\text{Cl}^- = \frac{8\cdot13}{6\cdot13 + 8\cdot13} \times 3\cdot14 = 1\cdot79 \text{ Å}$$

If we take structures containing univalent ions with inert gas structures, then by adding together the radii obtained by the Pauling method we get interionic distances that agree very closely with those measured experimentally by x-ray diffraction methods. These calculated values refer only to ions that are singly charged, however, and a correction factor has to be applied to get the radii of multivalent ions. The modified values, called 'crystal radii', are smaller than the Pauling 'univalent' radii, since the increased charge gives an increased force of attraction between the ions, pulling them closer together. The ionic radius also changes if the co-ordination number (the number of negative ions in contact with a given positive ion) changes. The values given in tables of crystal radii, some of which are quoted in *Table 10.3*, refer to structures in which the co-ordination number is six; these values have to be multiplied by a factor of 0·96 if the co-ordination number drops to four, and by a factor of 1·03 for ions with a co-ordination number of eight.

Table 10.3. Crystal Radii (Å)

Li+	Be²+		F⁻
0·60	0·31		1·36
Na+	Mg²+	Al³+	Cl⁻
0·95	0·65	0·50	1·81
K+			Br⁻
1·33			1·95
Rb+			I⁻
1·48			2·16

As we go down Group I of the Periodic Table, from lithium to rubidium, there is a steady increase in ionic size since the effect of the positive nuclear charge is counteracted by increased screening of the valency electron from the nucleus. There is, too, increase in the size of the negative ions as we go from fluoride to iodide. It should also be noted that negative ions are much larger than the corresponding isoelectronic positive ions (*cf.* F⁻ and Na+, Cl⁻ and K+), since the increased nuclear attraction in the positive ions pulls the electrons closer to the nucleus. Again, if we compare the radii of Na+, Mg²+ and Al³+, we observe a marked decrease in size; all the ions are isoelectronic, and the net increase in positive charge pulls the electrons closer to the nucleus.

IONIC STRUCTURES*

Since ions with noble gas electronic structures are spherical, their attraction for ions of opposite charge is exerted to the same extent in every direction, and simple electrostatic considerations show that in an ionic structure containing spherical A^+ and B^- ions, the most stable arrangement is one in which the ions are in contact with each other and arranged in a symmetrical way. The structures are determined by two factors: the relative sizes of the ions and the requirement that the structure, as a whole, must be electrically neutral. The size factor largely determines the geometry of the structure, and we can discuss this in terms of co-ordination numbers. The most stable arrangement for a co-ordination number of 2 is a linear one, B—A—B, because this minimizes the repulsion between the negatively charged B^- ions, and the stable arrangements for three, four, six and eight co-ordinations are co-planar, tetrahedral, octahedral and cubical, respectively. Now a large positive ion can be in contact with a large number of negative ions, but a small positive ion can only accommodate a small number of large negative ions around and in contact with it. Let us consider a particular example, where the co-ordination number of the A^+ ion is six.

Figure 10.2 (a) shows the immediate environment of an A^+ ion in the crystal structure; the A^+ ion is in contact with four co-planar B^- ions; there are, in addition, two more B^- ions in contact with A^+, one above and one below the plane of the diagram—these are omitted for clarity from *Figure 10.2*. *Figure 10.2 (b)* shows the corresponding environment of A^+ in another compound AC, where the radius of the C^- ion is greater than that of B^-; the negative ions are now in contact with each other as well as with the ion A^+. This represents a limiting condition for octahedral co-ordination.

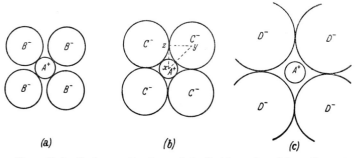

(a)   (b)   (c)

Figure 10.2. Ionic co-ordination and the limiting value of the radius ratio

* *Cf.* WELLS[8] and EVANS[9]

If a third substance AD were considered, containing a still larger negative ion D^-, it would be impossible to get a stable structure with six co-ordination, because the negative ions would then be in contact with each other but not with the positive ion *(Figure 10.2 (c))*. In this case, a more stable structure (with positive and negative ions in contact) would be obtained with fourfold co-ordination. The limiting condition, illustrated in *Figure 10.2 (b)*, is defined by the relative sizes of the ions; there is a limiting value of the radius ratio, r_+/r_-, which can easily be calculated from the geometry of *Figure 10.2 (b)*:

$$zy = xy \cos 45°$$

i.e.
$$r_- = (r_+ + r_-)/\sqrt{2}$$

whence
$$\frac{r_+}{r_-} = \sqrt{2} - 1 = 0.414$$

The radius ratio for the structure of the compound AD shown in *Figure 10.2 (c)* will be less than 0.414 since D^- is larger than C^-. Thus octahedral co-ordination is only possible if the radius ratio is greater than the limiting value. The limiting values of the radius ratio for other co-ordination numbers can easily be calculated and are given in *Table 10.4*. *Figure 10.3* is a three-dimensional representation of *(a)* anions in contact with each other but not with the cation, and *(b)* anions in contact with the cation.

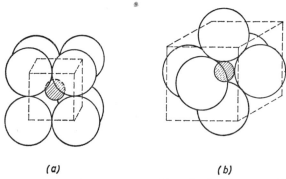

(a)   *(b)*

Figure 10.3. (a) Anions in contact with each other but not with the cation; (b) rock salt structure, six anions in contact with the cation

A knowledge of ionic radii may be a guide to the structure of an ionic crystal. Thus a radius ratio greater than 0.732 implies that from a geometrical point of view cubical co-ordination should be possible, whereas a ratio between 0.732 and 0.414 would require

Table 10.4. Limiting Values for the Radius Ratio

Co-ordination	Limiting value $\frac{r+}{r-}$
8 (*Cubic*)	0·732
6 (*Octahedral*)	0·414
4 (*Square-coplanar*)	0·414
4 (*Tetrahedral*)	0·225
3 (*Trigonal-coplanar*)	0·155

the co-ordination number to fall to 6 (octahedral). Unfortunately, the radius ratio is only a rough guide. Sodium fluoride has a radius ratio of 0·74, and potassium fluoride one of 1·0, but both these compounds crystallize in the rock salt structure with co-ordination numbers of 6. Although twelve fluoride ions can pack around a positive potassium ion, geometry does not permit a packing of twelve potassium around ions each fluoride ion in the crystal structure. Geometry would allow cubical co-ordination, however, giving a caesium chloride structure. It seems likely that in these cases 'second-order' effects such as 'van der Waals' forces may be decisive in determining the co-ordination, since there is only a small difference in crystal energy between a caesium chloride and a sodium chloride structure.

The co-ordination number of the positive ion is not necessarily the same as that of the negative ion. It will be the same in the case of structures of general formula AB, where the requirement of electrical neutrality means that there must be equal numbers of A^+ and B^- ions in the structure, but in a structure of formula AB_2, containing A^{2+} and B^- ions, there must be twice as many negative

Table 10.5. Ionic Structures

Type of structure	Examples	Radius ratio	Co-ordination number of positive ion	Co-ordination number of negative ion
Caesium chloride	CsCl, CsBr, CsI	>0·732	8 (*Cubic*)	8 (*Cubic*)
Rock salt	NaCl, NaBr, NaI } MgO, CaO, MnO }	<0·732 >0·414	6 (*Octahedral*)	6 (*Octahedral*)
Fluorite	CaF_2, SrF_2, ThO_2	>0·732	8 (*Cubic*)	4 (*Tetrahedral*)
Rutile	TiO_2, SnO_2, PbO_2	>0·414	6 (*Octahedral*)	3 (*Trigonal*)

as positive ions to preserve overall neutrality. The co-ordination of the positive ion must hence be double that of the negative ion.

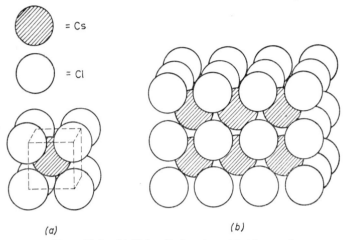

(a)

(b)

Figure 10.4. (a) Unit cell of caesium chloride crystal
(b) structure of caesium chloride

Table 10.5 summarizes the essential structural features of a number of simple inorganic substances. *Figure 10.4* illustrates the 'unit cell' of the caesium chloride structure and the way in which the units stack together to build up the crystal. The caesium ions form a simple cubic lattice interlocking with, but displaced from, a similar lattice of chloride ions. *Figure 10.5* shows the rock salt structure in which the sodium ions form a lattice with points at the corners of a cubic unit and also at the centres of the cube faces. The chloride ions form an interlocking face-centred cubic lattice.

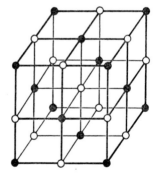

Figure 10.5. The rock salt structure

ELECTRONEGATIVITY

Although it is convenient to distinguish between covalent and ionic bonds, many bonds of a type intermediate between the purely ionic and the purely covalent are known. Thus, in the series of iso-electronic molecules CH_4, NH_3, OH_2 and FH, the electronic charge distribution in the C—H, N—H, O—H, and F—H bonds is increasingly concentrated in a direction away from the hydrogen atom, so that these essentially covalent bonds have increasing 'ionic character' as we go along the series from C—H to F—H. The C, N, O and F atoms are arranged in order of increasing 'electro-negativity'—defined as the power of an atom in a molecule to attract electrons to itself. This definition may be extended as follows:

Consider a diatomic molecule A—B: then, using valence-bond theory, we can describe this molecule by a wave function which is a linear combination of wave functions corresponding to structures such as A—B, A^+B^-, A^-B^+, i.e.

$$\psi = c_I \psi_I(A—B) + c_{II} \psi_{II}(A^+B^-) + c_{III} \psi_{III}(A^-B^+)$$

If, in this expression, $c_{III} > c_{II}$, we say that A is more electro-negative than B, whereas if $c_{III} = c_{II}$ the atoms have the same electronegativity.

There have been many attempts to assign quantitative electro-negativity values, and we shall only discuss the more important ones:

Method 1 (PAULING[7b])

If, in the expression for the wave function of the diatomic molecule A—B given above, $c_{II} = c_{III} = 0$, the bond is said to be purely covalent, and the bond energy, $D(A—B)$, will be equal to E_{cov}, the pure covalent bond energy. E_{cov} cannot be measured directly but its value is usually taken as the arithmetic mean of the bond energies of the covalent molecules A—A and B—B. If A and B have different electronegativities, however, the bond A—B will no longer be purely covalent, and the bond energy, $D(A—B)$, is greater than that of the purely covalent bond, E_{cov}, by an amount $\Delta(A—B)$, sometimes called the 'ionic-covalent resonance energy'.

Thus

$$D(A—B) = E_{cov} + \Delta(A—B)$$

or

$$\Delta(A—B) = D(A—B) - E_{cov}$$

Now the greater the difference between the electronegativities of A and B, the greater will be the value of $\Delta(A—B)$, and Pauling makes use of the following empirical expression to relate $\Delta(A—B)$ with electronegativity differences:

$$0.208 \sqrt{\Delta} = |\, x_A - x_B \,|$$

where x_A represents the electronegativity of atom A. (The factor 0.208 arises from the conversion of $\Delta(A—B)$ measured in kcal/mole into electron volt energy units.)

Pauling selected an arbitrary x value of 2.1 for hydrogen—this gave a suitable range of numerical values for the elements C to F in the first row of the Periodic Table; other values can then be assigned if the $\Delta(A—B)$ values are known. Unfortunately, however, the determination of $\Delta(A—B)$ is subject to considerable errors, and numerical values for electronegativities have to be scrutinized with some care.

Method 2 (Mulliken)

MULLIKEN[10] defines the electronegativity, M_A, of an atom A, as the arithmetic mean of its electron affinity, E_A, and its ionization potential I_A

i.e. $$M_A = \tfrac{1}{2}[E_A + I_A]$$

The reason for choosing this function can be explained qualitatively if we realize that, if A is more electronegative than B, then the energy required to form an ionic bond A^-B^+ from the neutral atoms will be less than that needed to form an ionic bond A^+B^-, since it is clearly easier to transfer the electron from atom B to atom A than vice versa, if atom A has the greater tendency to attract the electron. Now the energy which is required to produce A^- and B^+ from A and B is given by $I_B - E_A$; similarly, the energy needed to produce A^+ and B^- is given by $I_A - E_B$. Thus if

$$I_A - E_B = I_B - E_A$$

then atoms A and B must be equally electronegative. On the other hand, A will be more electronegative than B if

$$I_A - E_B > I_B - E_A$$

i.e. if

$$I_A + E_A > I_B + E_B$$

The sum of the ionization potential and electron affinity is thus a measure of the electronegativity. It is important to note, however, that the appropriate values of ionization potential and electron affinity are not the same as those of an isolated atom. Electronegativity relates to an atom combined with other atoms, so that in beryllium, for example, where the atom uses sp hybrid orbitals, we can obtain two different ionization potentials, corresponding to the removal of the s or the p electron. We saw on page 51 that s electrons are more firmly held than p electrons of the same principal quantum number, so that the value of ionization potential used in electronegativity determinations will depend upon the nature of the bond hybridization (see also page 125). The Mulliken approach is also limited by the difficulty of getting reliable electron affinity values. A plot of Mulliken electronegativities against Pauling values gives a straight line going through the origin, and the slope of this line gives the equation

$$x_{\text{Pauling}} = x_{\text{Mulliken}}/3 \cdot 15$$

relating the two electronegativity scales (PRITCHARD and SKINNER[1]).

Method 3 (Sanderson)

R. T. SANDERSON[11] suggests that electronegativity is related to the 'compactness' of the electron charge cloud of an atom relative to that of a hypothetical isoelectronic inert atom, since if an atom is very electronegative, *i.e.* strongly attracts other electrons, its own electrons will be held close together. He therefore represents electronegativities as a Stability Ratio, SR, defined as the ratio of the average electron density of the atom to that of a hypothetical isoelectronic inert atom. The Sanderson and Pauling values are related by the expression

$$\sqrt{x_{\text{Pauling}}} = 0 \cdot 21 \text{ SR} + 0 \cdot 77$$

The two scales agree quite closely except for germanium, arsenic and antimony, where the Sanderson values are appreciably higher than the corresponding Pauling electronegativities. A high value for Ge is also obtained in an electronegativity scale (ALLRED and ROCHOW[12]) based on calculations of the electrostatic force exerted by a nucleus on an electron in a bonded atom. Sanderson, and Allred and Rochow, claim that there is a great deal of chemical evidence in favour of the high value for germanium; DRAGO[13], however, has pointed out that these chemical properties can equally be explained, while retaining the Pauling values for electronegativity, as a consequence of the presence of d electrons in the electron configuration of the germanium atom.

PRITCHARD and SKINNER[1] have reviewed many suggested electro-negativity scales, and *Table 10.6* lists a selection of the values they recommend. This concept of electronegativity is also of importance to theoretical organic chemistry, where chemical reactivity can be correlated with differences in the electron charge density around particular atoms. The ionization potential of an *s* electron is greater than that of a *p* electron, since the *s* electron is, on the average, more under the influence of the nucleus. This means that in a hybrid (*sp*) orbital, the greater the *s* character of the hybrid the greater will be the effective electronegativity of the atom giving rise to the hybrid orbital. The electronegativity of the carbon atom in acetylene (*sp* hybrids) is thus greater than it is in methane, where carbon is using sp^3 hybrid orbitals. It is this fact that accounts for the acidic properties of acetylene, *e.g.* the ease with which one of its hydrogen atoms can be replaced by sodium.

THE HYDROGEN BOND*

The covalent and the ionic bonds that we have been discussing are usually very strong, with bond energies in the range *ca.* 30–130 kcal/mole. We now discuss a very much weaker link, the so-called 'hydrogen bond', where the bond energies are in the 2–10 kcal/mole range. A hydrogen bond between two atoms *A* and *B* can be written *A*—H ... *B*; this implies that *A* and *B* are sufficiently close together for a bond to exist, and that the hydrogen atom involved in the bond is attached to atom *A* by a covalent bond. Hydrogen bonding is only important when the atoms *A* and *B*

Table 10·6.   Electronegativity Values

H 2·1						
Li 1·0	Be 1·5	B 2·0	C 2·5	N 3·0	O 3·5	F 3·9
Na 0·9	Mg 1·2	Al 1·5	Si 1·8	P 2·1	S 2·5	Cl 3·0
K 0·8	Ca 1·0	—	Ge 1·8	As 2·0	Se 2·4	Br 2·8
Rb 0·8	Sr 1·0	—	Sn 1·8	Sb 1·9	Te 2·1	I 2·5
Cs 0·7	Ba 0·9					

* PIMENTEL and McCLELLAN[14]

140

Table 10.7.   Hydrogen Bond Lengths

A	B	Example	$R_{A-H \ldots B}$ (Å)
F	F	HF_2^- in KHF_2	2·26
O	O	HCO_3^- in $KHCO_3$	2·61
		H_3BO_3	2·70–2·73
		Ice (cubic)	2·76
N	O	Proteins	2·67–3·07

are strongly electronegative—e.g. nitrogen, oxygen and fluorine, and Table 10.7 lists the lengths of some typical hydrogen bonds in selected compounds of these elements.

Hydrogen Bonding and Crystal Structure

The structures of many crystalline substances are determined by hydrogen bonding, and we can conveniently discuss these as structures in which there are (a) discrete complex ions containing hydrogen bonds (b) ions linked through hydrogen bonds into infinite chains (c) infinite two-dimensional layers and (d) three-dimensional macro-molecules.

(a) Discrete ions—A detailed analysis of KHF_2 by neutron diffraction reveals an ionic crystal containing K^+ cations and HF_2^- anions. The anion is linear, the two fluorine atoms being linked through a hydrogen atom mid-way between them, i.e. F—H .. F.

(b) Chain structures—The planar carbonate ions in $KHCO_3$ are linked through hydrogen bonds into infinite chains:

Hydrogen bonding is also an essential feature of protein structures, where the molecules are built up by the linking of peptide (amide) units to form long chains. There is now considerable evidence that

in many crystalline proteins these polypeptide chains are coiled into a spiral structure, the a-helix, held together by hydrogen bonds linking the carbonyl oxygen of one peptide residue with a nitrogen atom in the third unit along the chain.

(c) Infinite layers—Boric acid is a good example of the way in which a planar group of atoms such as $(BO_3)^{3-}$ can be linked into a planar network through hydrogen bonding. In the structure shown dashed lines indicate the position of the hydrogen bonds.

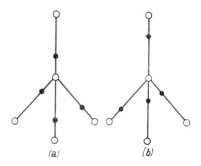

(d) Three-dimensional macromolecules—Ice can exist in a number of crystalline forms[15]. Thus a cubic (diamond) structure can be obtained below − 80° C, while at 0° C ice has a hexagonal structure. In the cubic form, each oxygen atom is surrounded tetrahedrally by four other oxygen atoms at a distance of 2·76 Å, giving a very open structure which owes its stability to hydrogen bonding. The hydrogen atoms are located between the oxygen atoms in such a way that, at any instant, one oxygen atom has two hydrogen atoms attached at distances corresponding with the length of a covalent O—H bond (0·99 Å) and two hydrogen atoms at much

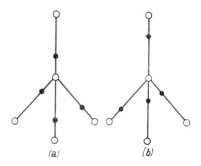

Figure 10.6. The structure of ice: *(a)* and *(b)* show alternative arrangements of the hydrogen atoms (represented by small black circles)

greater distances (1·77 Å) held by hydrogen bonding *(Figure 10.6 (a))*.

At another instant the hydrogen atoms may have the configuration indicated in *Figure 10.6 (b)*. A hydrogen atom does not therefore have a precise location between two oxygen atoms in the ice crystal; it effectively oscillates between two extreme positions.

Theories of Hydrogen Bonding

A hydrogen atom, of configuration $1s^1$, can form only one covalent bond, since the $2s$ and $2p$ orbitals have energies too high to be involved in the formation of additional bonds. We have, therefore, to seek another explanation of the way in which hydrogen can link two atoms together in the bond A—H ... B.

(a) The electrostatic approach—The fact that hydrogen bonding occurs only between very electronegative elements suggests that electrostatic interaction between polar groups may be a sufficient explanation.

Let us consider the covalent bond between hydrogen and some very electronegative atom A. The electron charge cloud between the two atoms will be considerably distorted to give a greater concentration of electron density in the vicinity of A and a strong dipole will be produced. A second electronegative atom B attached to another atom or molecule, will also form the negative end of a dipole. If the two dipoles approach one another along the line A—H ... B, then the electrostatic attraction between the positive end of the dipole on A—H and the negative charge on B will be greater than the repulsive forces between like charges, and the closer that B can approach to H the stronger will be this electrostatic link between them. The strength of this electrostatic attraction can be calculated approximately, assuming a model in which orbitals are represented by point charges at their centroids. Lonepair electrons play an important part in determining the strength and direction of hydrogen bonds, and SCHNEIDER's[16] calculations on the water molecule indicate that the strongest bond is formed when the O—H axis of one molecule is collinear with the lone-pair axis on the oxygen atom of a neighbouring water molecule. Calculations of this kind usually give a reasonable hydrogen bond energy of about 6 kcal/mole, but there are some objections to this theory (*see*, especially, COULSON[17]) and it seems that in many cases the hydrogen bond may have some covalent character.

(b) Valence-bond treatment—COULSON and DANIELSSON[18] have at-

tempted to determine the covalent contribution to the hydrogen bond by considering structures such as

$$A—H \qquad B \qquad \text{covalent } A—H \text{ bond}$$
$$A^- \ H^+ \dots B \qquad \text{ionic } A—H \text{ bond}$$
$$A^- \ H————B^+ \qquad \text{covalent } H—B \text{ bond}$$

and a later treatment by TSUBOMURA[19] considers also the structure

$$A^+ \ H^- \dots B$$

The calculations are necessarily approximate, and they involve a considerable number of assumptions but they indicate that the covalent contribution may be appreciable, especially in 'short' hydrogen bonds (*ca.* 2·5 Å).

(c) Molecular-orbital treatment—PIMENTEL[20] has discussed hydrogen bonding in the HF_2^- ion using molecular-orbital theory. He takes linear combinations of the hydrogen s orbital with fluorine p orbitals directed along the bond. Thus, if the p orbitals of the two fluorine atoms are denoted p_A and p_B, the symmetrical F—H ... F bond is described by

$$\psi_1 \text{ (bonding)} = p_A + p_B + a_1 s$$
$$\psi_2 \text{ (non-bonding)} = p_A + p_B$$
$$\psi_3 \text{ (anti-bonding)} = p_A + p_B - a_3 s$$

where a_1 and a_3 are mixing coefficients. The two electrons involved in bond formation are contained in a molecular orbital (ψ_1) which spreads out axially on either side of the hydrogen atom, thus forming two equivalent weak bonds, and there are also two electrons concentrated mainly on the fluorine atoms in the non-bonding orbital ψ_2.

The Importance of Hydrogen Bonding

The comparatively small value of the hydrogen bond energy has some important consequences, because chemical reactions which involve the breaking of weak bonds may take place readily at room temperature; this is especially the case in many biochemical systems.

Hydrogen bonding is also of importance in explaining the properties of certain liquids; *e.g.* the high dielectric constants, high boiling points, and low vapour pressures of solvents such as water and the alcohols, the so-called polar solvents, are ascribed to intermolecular association through hydrogen bonds. Intramolecular (or internal) bonding may also occur; thus in *o*-nitro-phenol the hydrogen of the phenolic group and an oxygen atom of the nitro group are sufficiently near to each other for appreciable

electrostatic interaction to occur, and a six-membered ring system is formed. This explains why the *o*-isomer is more volatile than the *m*- and *p*-nitrophenols; the latter show intermolecular associa-tion and have low vapour pressures, whereas the *o*-isomer is not associated since its hydrogen bond has been formed internally, and the isomers can be readily separated from each other.

METALLIC BONDS

Metal structures have some very characteristic properties. Each atom in the crystal has a very high co-ordination number (frequently twelve but sometimes eight), and the structure has high electrical and thermal conductivity. The metal atoms are packed close together in the crystal; this implies that there is extensive overlap of the outer electron orbitals and that the valency electrons can no longer be associated with a particular nucleus, but are completely delocalized over all the atoms in the structure. A given metal can thus be regarded as an assembly of positive ions which, in general, are spheres of identical radius (manganese and uranium are excep-tions) packed together to fill space as completely as possible. There are two ways of 'close-packing' identical spheres; one of these has hexagonal and the other cubic symmetry, but in each case the co-ordination number is twelve. The 'body-centred' cubic structure of, *e.g.*, the alkali metals is less closely packed, for here the co-ordination number is eight, and each ion has eight nearest neigh-bours at the corners of a surrounding cube.

Valence-Bond Theory of Metals

The bonds between one atom in a metal and its nearest neigh-bours cannot be electron-pair links, since there are not enough electrons available—an alkali metal atom, for example, has only one electron to share with eight neighbours. However, we can write down a large number of structures of the type

145

$$Na\text{---}Na \quad Na\text{---}Na \qquad\qquad Na\text{---}Na \quad Na \quad Na$$
$$\text{I} \qquad\qquad\qquad\qquad | \quad\quad | \qquad \text{II}$$
$$Na\text{---}Na \quad Na\text{---}Na \qquad\qquad Na\text{---}Na \quad Na \quad Na$$

and compound them to give a structure which is a resonance hybrid of these forms. The bond will be weak, however, since one electron pair has to link eight atoms. A considerably stronger bond will be obtained if, as PAULING[7b] suggests, we also consider structures such as

$$Na^{+} \quad Na\text{---}Na \quad Na$$
$$|$$
$$Na\text{---}Na^{-}\text{---}Na \quad Na$$

in which there are negative ions forming two covalent bonds by making use of sp hybrid orbitals. These structures only become important in the linear combination if there are unoccupied orbitals of suitable energy available for hybridization. If these orbitals are available, there will be a large number of these structures, and the bond will be correspondingly stronger. Thus in sodium the energy of the $3p$ orbitals, normally unoccupied in isolated sodium atoms, is not much greater than that of the $3s$ orbital containing the valency electron when the atoms are close together as they are in the metal, and sp hybrid orbitals can be formed. Carbon, on the other hand, is a non-metal, since, in the diamond structure, the four tetrahedrally arranged sp^3 hybrid orbitals are all doubly occupied, and the energies of the $3s$ and $3p$ orbitals are too great to be available for the formation of new hybrid orbitals.

The strength of the bond in these structures will depend upon the number of valency electrons that each atom can contribute. Thus, as we go along the first long row of the Periodic Table, the K, Ca, Sc, Ti, V and Cr atoms can contribute 1, 2, 3, 4, 5 and 6 electrons, respectively, to the structure, and the corresponding increase in bond strength from K to Cr is made evident by the steady increase in melting points and hardness, and a decrease in interatomic distances. These physical properties remain roughly constant from Mn to Ni, and Pauling assigns a 'metallic' valency of 6 to these elements also. His values of $5\frac{1}{2}$ and $4\frac{1}{2}$ for the 'metallic' valencies of copper and zinc, respectively, are based on magnetic properties of these elements, the fractional values representing situations in which, at a given instant, some atoms are in one valency state (*e.g.* 6) and some in another (*e.g.* 3 or 4). These values have, however, been criticized (*see, e.g.,* HUME-ROTHERY[21] for a critical analysis of Pauling's treatment).

146

Hybrid Orbitals in Metal Structures

ALTMANN, COULSON and HUME-ROTHERY[22] have likewise discussed metal structures in terms of the directional character of hybrid orbitals. Just as carbon atoms in the diamond are arranged in a structure determined by the tetrahedral disposition of sp^3 hybrid orbitals, so, in a metal, hybrid orbitals may also determine the structure. In metals, however, the hybrid orbitals are only 'partially occupied'. This means that we can write down a large number of possible structures in some of which a particular orbital may be occupied by electrons, whereas this same orbital may be unoccupied in other structures. The linear combination of wave functions corresponding to these structures then describes a state in which the hybrid orbitals are 'partially occupied'.

The details of their discussion lie outside the scope of this book but, briefly, they consider that body-centred metal structures are determined by the geometry of sd^3 hybrid orbitals, cubic close-packed structures by p^3d^3 hybrids and hexagonal close-packed structures by sd^2, pd^5 and spd^4 hybrid orbitals. These hybrids have differing d 'character' or 'weight' and the structure adopted by a transitional metal, for example, is determined by the number of d electrons it can contribute to these hybrid orbitals.

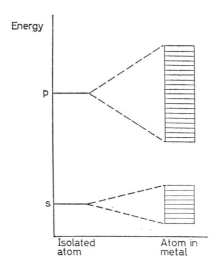

Figure 10.7. Energy levels and energy bands in isolated atoms and in metal structures

Molecular-Orbital Theory of Metals

The delocalization of the 'free' electron orbitals over all the atoms of a metal structure is, essentially, a molecular-orbital approach to metallic bonding. Now we saw in Chapter 7 that when two atoms combine to form a diatomic molecule, two molecular orbitals are formed by the overlap of atomic orbitals of the combining atoms. When a third atom is added to the diatomic molecule, three molecular orbitals are obtained and, in general, when there is overlap of orbitals on N atoms in a solid structure, N molecular orbitals will be obtained. Now each orbital has an associated energy value and *Figure 10.7* illustrates diagrammatically the difference between the energy levels of two orbitals of an isolated atom, and the large number of closely spaced levels which result from the overlapping of orbitals in a metal structure. The levels become so close together in the metal that they effectively form a 'band' of energy values. These bands may be separated from each other, as in *Figure 10.7*, or they may overlap. Each level can accommodate two electrons, so that, if there are N 'free' electrons, the first $N/2$ states will be doubly-occupied.

Now electrical conduction depends upon the movement of electrons through the structure under the influence of an applied field, and this movement can only occur if the electrons can accept energy and move to higher, unoccupied levels. Metals, therefore, are structures in which accessible, unoccupied levels are available, whereas in non-metallic substances, which are usually insulators, all the accessible levels are already occupied. The so-called 'semiconductors' are substances in which there is an unoccupied energy band not too far removed from a band which is completely occupied. At low temperatures the electrons in the occupied band have not got sufficient energy to transfer to the unoccupied or 'conduction' band, but at high temperatures they may acquire sufficient energy to make this transition; the increase in conductivity with temperature, characteristic of semi-conductors, is thus explained.

THE MOLECULAR OR VAN DER WAALS BOND

The 'non-ideal' behaviour of gases on compression was ascribed by van der Waals to the existence of weak forces of attraction between atoms or molecules in the gaseous state. These forces are also present in the liquid and solid state; thus crystals of the noble gases have structures in which the monatomic molecules are held in a close-packed arrangement by the van der Waals forces. The same force produces the so-called 'molecular bond' linking,

148

e.g. discrete diatomic molecules in solid halogen structures. Many organic crystals, *e.g.* naphthalene, have structures in which discrete molecules are linked by molecular bonds, and the low melting points of such crystals show that the link is very weak compared with covalent or electrovalent bonds—about 10 kcal/ mole.

A quantum theory of this bond has been worked out by LONDON[23] but an account of it is outside the scope of this book.

REFERENCES

[1] PRITCHARD, H. O. and SKINNER, H. A.   *Chem. Rev.*   55 (1955) 745
[2] WADDINGTON, T. H.   *Adv. Inorg. Chem. Radiochem.*   1 (1959) 157
[3] BAYLISS, N. S.   *Q. Rev. chem. Soc.*   6 (1952) 319
[4] WALSH, A. D.   *Q. Rev. chem. Soc.*   2 (1948) 73
[5] WHEATLEY, P. J.   *The Determination of Molecular Structure*, Clarendon Press, Oxford, 1959
[6] LONSDALE, K.   *Crystals and X-rays*, Bell, London, 1948
[7] PAULING, L.   *The Nature of the Chemical Bond*, Cornell Univ. Press, New York, *3rd Edition*, 1960 (*a*) 511; (*b*) 88
[8] WELLS, A. F.   *Structural Inorganic Chemistry*, Clarendon Press, Oxford, *3rd Edition*, 1962
[9] EVANS, R. C.   *An Introduction to Crystal Chemistry*, Cambridge Univ. Press, *2nd Edition*, London, 1964
[10] MULLIKEN, R. S.   *J. chem. Phys.*   2 (1934) 782; *see also* MOFFITT, W. E. *Proc. R. Soc.*   A 196 (1949) 510
[11] SANDERSON, R. T.   *J. chem. Phys.*   23 (1955) 2467
[12] ALLRED, A. L. and ROCHOW, E. G.   *J. inorg. nucl. Chem.*   5 (1958) 264
[13] DRAGO, R. S.   *J. inorg. nucl. Chem.*   15 (1960) 237
[14] PIMENTEL, G. C. and McCLELLAN, A. L.   *The Hydrogen Bond*, Freeman, San Francisco and London, 1960
[15] LONSDALE, K.   *Proc. R. Soc.*   A 247 (1958) 424
[16] SCHNEIDER, W. G.   *J. chem. Phys.*   23 (1955) 26
[17] COULSON, C. A.   *Research, Lond.*   10 (1957) 149
[18] COULSON, C. A. and DANIELSSON, U.   *Ark. Fys.*   8 (1954) 239
[19] TSUBOMURA, H.   *Bull. chem. Soc. Japan*   27 (1954) 445
[20] PIMENTEL, G. C.   *J. chem. Phys.*   19 (1951) 446
[21] HUME-ROTHERY, W.   *A. Rep. chem. Soc.*   46 (1949) 42
[22] ALTMANN, S. L., COULSON, C. A. and HUME-ROTHERY, W.   *Proc. R. Soc.* A 240 (1957) 145
[23] LONDON, F.   *Trans. Faraday Soc.*   33 (1937) 8

III

THE APPLICATION OF THE PRINCIPLES
OF CHEMICAL BONDING

11

THE STRUCTURES OF SOME SIMPLE
INORGANIC COMPOUNDS

BOND LENGTHS

WE have quoted a number of bond lengths in previous chapters, and
have pointed out that multiple bonds are shorter than single bonds
for a given pair of atoms. We have also seen that even if there is
no multiple bonding, the length of a given bond may not always be
the same in different compounds, but may depend on the types of
orbitals used by the linked atoms; thus, in Chapter 9, we saw how
the C—X bond length varied with the s character of the hybrid
orbitals used by the carbon atoms. It is possible to assign values
for single (and sometimes multiple) bond lengths (bearing in mind
the limitations imposed by hybridization), and *Table 11.1* lists a
selection of such values, based on the Chemical Society's *Table of
Interatomic Distances (1958, 1965)*[1] together with some more recent
measurements. It should, however, be noted that whereas, for
example, C—C bond lengths are obtained from a statistical analysis
of a large number of experimental measurements, bond lengths in
inorganic compounds are based usually on relatively few measurements.

CALCULATED BOND LENGTHS

We shall from time to time refer to calculated values for some bond
lengths. These values are obtained by adding the 'covalent radii'
for the linked atoms, where the covalent radius of a given atom A is
obtained by taking one-half of the measured A—A bond length.
Thus the single-bond covalent radius of carbon is 0·77 Å since the
C—C single-bond length is 1·54 Å. The single covalent radius for
chlorine is 0·99 Å (one-half of the Cl—Cl bond length), so that the
length of the C—Cl bond is calculated to be 1·76 Å (0·77 + 0·99);
the experimental value for the C—Cl distance in carbon tetrachloride is 1·77 Å. *Table 11.2* gives values for the covalent radii
for some of the more important elements that we shall be discussing
in this chapter. *Table 11.3* lists values for the covalent radii of
some elements forming multiple bonds.

Comparison of experimentally measured bond lengths *(Table*

151

Table 11.1.   Selected Bond Lengths

Bond	Bond length (Å)	Compound	Hybridization
B—Br	1·87	BBr_3	sp^2
B—C	1·56	$B(CH_3)_3$	sp^2
B—Cl	1·75	BCl_3	sp^2
B—F	1·30	BF_3	sp^2
B—O	1·36	$B(OH)_3$	sp^2
C—Br	1·94	CBr_4	sp^3
C—C	1·54	C_2H_6	sp^3
C=C	1·33	C_2H_4	sp^2
C≡C	1·20	C_2H_2	sp
C—Cl	1·77	CCl_4	sp^3
C—F	1·32	CF_4	sp^3
C—H	1·09	CH_4	sp^3
C—N	1·47	$CH_3 . NH_2$	(sp^3)
C≡N	1·16	HCN	sp
C—O	1·42	$(CH_3)_2O$	(sp^3)
C=O	1·22	$(CH_3)_2CO$	(sp^2)
C—S	1·82	$(CH_3)_2S$	(sp^3)
Si—Br	2·15	$SiBr_4$	sp^3
Si—C	1·89	$Si(CH_3)_4$	sp^3
Si—Cl	2·01	$SiCl_4$	sp^3
Si—F	1·55	SiF_4	sp^3
Si—H	1·48	SiH_4	sp^3
Ge—Br	2·29	$GeBr_4$	sp^3
Ge—C	1·98	$Ge(CH_3)_4$	sp^3
Ge—Cl	2·08	$GeCl_4$	sp^3
Ge—F	1·67	GeF_4	sp^3
Ge—H	1·53	GeH_4	sp^3
Sn—Br	2·44	$SnBr_4$	sp^3
Sn—C	2·18	$Sn(CH_3)_4$	sp^3
Sn—Cl	2·31	$SnCl_4$	sp^3
Sn—H	1·70	SnH_4	sp^3
N—C	1·47	$N(CH_3)_3$	(sp^3)
N—F	1·37	NF_3	(sp^3)
N—H	1·01	NH_3	(sp^3)
P—C	1·84	$P(CH_3)_3$	(sp^3)
P—Cl	2·04	PCl_3	(sp^3)
P—F	1·54	PF_3	(sp^3)
P—H	1·44	PH_3	(sp^3)
O—H	0·96	OH_2	(sp^3)
S—Cl	1·99	SCl_2	(sp^3)
S—F	1·58	SF_6	sp^3d^2
S—H	1·33	SH_2	(sp^3)

(i) Values for boron based on sp^2 hybridization.

(ii) Values for carbon and other Group IV elements based on sp^2 hybridization (except for double and triple bonds—which are sp^2 and sp respectively).

(iii) Groups V and VI. Here the values refer only to the bond in the example quoted; the N—F distance in NF_3 is not the same as that in N_2F_2 or FNO. The elements form many bonds which appear to have multiple bond character (*e.g.* N—O, P—O, S—O); these bonds are excluded from the table, but they are discussed later in this chapter.

(iv) (sp^3) indicates a modification of sp^3 hybridization arising either from the non-equivalence of the four attached groups, or because fewer than four groups are bonded and lone-pair electrons occupy bonding positions.

11.1) with those calculated using single bond radii from *Tables 11.2* and *11.3* shows that the calculated values are almost always too high. This is because these values may relate to bonds between atoms of differing electronegativity, whereas the covalent radii are

Table 11.2. Covalent Radii (Å) (Single Bonds)

H						
0·37						
Li	Be	B	C	N	O	F
1·23	0·89	0·80	0·77	0·74	0·74	0·72
		Al	Si	P	S	Cl
		1·25	1·17	1·10	1·04	0·99
		Ge	As	Se	Br	
		1·22	1·21	1·17	1·14	
		Sn	Sb	Te	I	
		1·41	1·41	1·37	1·33	

Table 11.3. Covalent Radii (Å) (Multiple Bonds)

	C	N	O	S	Se
Double Bond	0·67	0·60	0·55	0·94	1·07
Triple Bond	0·60	0·55	0·50		

obtained from bond lengths in homonuclear molecules. A corrected bond length can be calculated in these cases by making use of a relationship proposed by SCHOMAKER and STEVENSON[2]:

$$r_{A-B} = r_A + r_B - 0·09\, \Delta$$

where r_{A-B} is the bond length, r_A and r_B the covalent radii for atoms A and B, respectively, and Δ is the difference between the electronegativities of A and B. It must be emphasized, however, that these summations of covalent radii only give approximate bond lengths and do not allow for small variations arising from changes in hybridization.

BOND CONFIGURATION FOR FIRST ROW ELEMENTS

In Chapter 9 we discussed the linear, trigonal-planar and tetra-hedral arrangement of bonds used by elements in the first row of the Periodic Table, and we saw that all three configurations were to be found in carbon compounds. Thus we get the tetrahedral arrangement, as in methane, if all four valency electrons are used for σ-bonding, while structures with trigonal-planar or with linear symmetry arise when only three or two σ bonds, respectively, are formed:

$$\underset{\underset{H \quad H \quad H}{}}{\overset{H}{\underset{|}{C}}} \qquad \underset{H}{\overset{H}{}}C{=}C\underset{H}{\overset{H}{}} \qquad H{-}C{\equiv}C{-}H$$

(or CCl_4)   (or $\underset{Cl}{\overset{Cl}{}}C{=}O$)   (or $O{=}C{=}O$)

The structural 'skeleton' of these molecules is determined by the σ bonds.

The same three basic configurations are found in compounds of some other first row elements. Thus the ions N^+, B^- and Be^{2-} have the same number of electrons as carbon (i.e. they are 'isoelectronic' with carbon) and will give similar structures, e.g. the tetrahedral ions $[NH_4]^+$, $[BF_4]^-$ and $[BeF_4]^{2-}$. If only three σ bonds are formed we get the planar structure, so that the nitrate ion (considering one of the valence-bond forms) has a structure analogous to that of phosgene:

$$\underset{O^-}{\overset{O^-}{}}N^+{=}O \quad \text{and} \quad \underset{Cl}{\overset{Cl}{}}C{=}O$$

Now we saw in Chapter 9 (page 114) that we could discuss ammonia and water molecules in terms of tetrahedral configurations, although the bond angles differed from the tetrahedral value of 109·5°, because some bonding positions were occupied by lone-pair electrons. These deviations arise from the fact that the charge clouds of the bonding electron pairs are concentrated around the line joining the linked nuclei, whereas the lone-pair orbital can 'spread out' and force the bonding pair orbitals closer together. These ideas have been extended by GILLESPIE and NYHOLM[3] to cover a wide range of compounds. Thus the observed bond angles can be explained if it is assumed that repulsion between electron pairs decreases in the order

lone pair—lone pair > lone pair—bonded pair >
bonded pair—bonded pair

We shall be discussing this approach in more detail later in the chapter.

Oxygen has two lone pairs and two bonding pairs in the water molecule, and once again the basic structure is tetrahedral, although the presence of two lone pairs results in a contraction of the HOH angle to 104·5°. A negatively charged nitrogen atom is isoelectronic with oxygen, so that the amide ion $[NH_2]^-$, has a structure similar to that of water. Fluorine forms only one σ bond, and retains three lone pairs, but the structure can still be considered basically tetrahedral, and indeed we shall see later that this approach helps us to understand the nature of the hydrogen bonding in the zig-zag polymers $(HF)_n$.

We can therefore consider all compounds of carbon, nitrogen, oxygen, and fluorine which contain only σ bonds, as tetrahedral molecules in which a lone pair of electrons progressively replaces a bonding pair.

Boron and beryllium have insufficient electrons to provide the tetrahedral configuration, so in their simple compounds only a lower degree of symmetry (planar and linear, respectively) is attained. Thus BCl_3 may be compared with $COCl_2$, and BeX_2 with CO_2.

Table 11.4. Configurations for Elements of the First Row

Linear	Planar (trigonal)	Tetrahedral	Tetrahedral (1 lone pair)	Tetrahedral (2 lone pairs)	Tetrahedral (3 lone pairs)
—Be—	B	Be^{2-}	N*	N^-	O^-
—C≡	C	B^-	O^+	O	F
=C=	N^+	C			
		N^+			

* The dotted line will always be used to represent a lone pair of electrons.

155

Boron and beryllium compounds of planar and linear symmetry will accept electrons from donor molecules (those with lone pairs), and form compounds such as $BF_3.NH_3$ and $BeCl_2.2OEt_2$ in which an approximately tetrahedral arrangement is attained. *Table 11.4* illustrates how the whole structural chemistry of σ-bonded compounds of the first row elements can be systematized in terms of these three basic configurations.

The maximum number of bonds that can be formed by elements of the second and third rows increases from four to six and eight, respectively, so that with these heavier elements we must expect other configurations besides those found for compounds of the first row elements to be important. In particular we get a trigonal-bipyramidal arrangement for five electron pairs, and octahedral structures for six electron pairs which will be discussed in some detail later in this chapter, and again in Chapter 12.

We now review the compounds formed by the non-transitional elements, bearing in mind the basic configurations we should expect to find, and the modifications that may arise through inter-electronic repulsions.

LITHIUM AND THE ALKALI METALS

Each of these elements has one lone electron in an outer *s* orbital, and this electron is readily lost to give the unipositive metal ion. The compounds of these metals are accordingly almost always ionic in character, and many of their properties, *e.g.* hardness, solubility, hydration, can be related to the factors discussed in Chapter 10, the ease of formation, the stability, and the size of the ions. The crystalline structures are ordered arrangements of ions and have comparatively high melting points; the molten compounds are good conductors of electricity, and even in the gaseous phase 'ion pairs' can be detected. Such substances are usually soluble in polar solvents (*e.g.* H_2O, NH_3), in which they are almost completely dissociated in dilute solution. These properties are illustrated by typical salts such as lithium chloride and sodium sulphate.

Two of the more interesting examples of these ionic substances are the hydrides and oxides. Lithium hydride (LiH), for instance, is formed as a colourless, crystalline solid by the direct union of the elements at 700°–800°C; it may be heated to 1000°C before it decomposes. Electrolysis of the solid at a temperature just below the melting point liberates lithium at the cathode and hydrogen at the anode (PETERS[4]), thus indicating the presence of Li^+ and H^- ions in the substance. Lithium hydride has the rock salt structure *(Figure 10.5)*, page 136). Direct oxidation of the alkali metals gives, as main products, the compounds Li_2O, Na_2O_2, KO_2, RbO_2, CsO_2; these are all ionic crystals, but whereas the anion in Li_2O is the simple oxide ion O^{2-}, the anions in Na_2O_2 and KO_2 are the peroxide ion, O_2^{2-}, and the superoxide ion, O_2^-, respectively. The O_2^- ion contains one unpaired electron, and it is accordingly paramagnetic with the configuration

$$(KK)(z\sigma)^2(y\sigma)^2(x\sigma)^2(w\pi)^4(v\pi)^3$$

Although these alkali metals form predominantly ionic compounds, they can form covalent molecules such as Li_2, Na_2, . . . *etc.*, by pairing their outer s electrons. These diatomic alkali metal molecules are found in small amounts (~ 1 per cent) in the metal vapour; their dissociation energies ($Li_2 = 25$, $Na_2 = 17\cdot3$, $K_2 = 11\cdot8$, $Rb_2 = 10\cdot8$, $Cs_2 = 10\cdot4$ kcal/mole) show that the bond becomes weaker as the atoms become larger. The alkali metal alkyls show an interesting variation in properties. Alkyls of sodium and heavier metals are colourless, involatile solids that are insoluble in benzene and react violently in air; these properties indicate an ionic structure, M^+R^-, the intense reactivity arising through the localized negative charge. The lithium analogues are much less reactive, however, and (except for the methyl) dissolve in hydrocarbon solvents to give tetrameric or hexameric species (BROWN et al.[5], WEINER et al.[6]). Recent x-ray investigations on lithium methyl (WEISS and LUCKEN[7]) and ethyl (DIETRICH[8]) have indicated tetrameric units; in the case of the methyl, the lithium atoms form a tetrahedral arrangement with methyl groups at the centres of each face. It is clear that the bonding in such compounds is not ionic, and we shall see in Chapter 13 that the best description involves delocalized σ orbitals that embrace an alkyl group and two or more lithium atoms.

BERYLLIUM AND THE ALKALINE EARTH METALS

Because of the increased effective nuclear charge, which 'pulls in' the outer electron charge cloud, the beryllium atom is much smaller than the lithium atom (radii: Be = $0\cdot89$ Å, Li = $1\cdot23$ Å) and its ionization potential is greater (first ionization potentials Be = $214\cdot8$ kcal/mole, Li = $124\cdot2$ kcal/mole; second ionization potential Be = 420 kcal/mole). This higher ionization potential of beryllium is reflected in the greater tendency for beryllium to form covalent compounds. As we go down the group from beryllium to barium, the atomic size increases, and the ionization potential decreases *(see Table 11.5)*, so that we should expect a corresponding increase in ionic characteristics. Beryllium, indeed, by virtue of its size, occupies a unique position in the group; it forms mainly covalent

Table 11.5. Atomic Radii and First Ionization Potentials for the Group IIA Elements

Element	Atomic radii (Å)	1st ionization potentials (kcal/mole)
Be	0·89	214·8
Mg	1·36	175·3
Ca	1·74	140·3
Sr	1·92	130·6
Ba	1·98	119·6

compounds. Compounds formed by the remaining Group IIA elements are almost always ionic.

Electrolytic oxidation of aqueous solutions of beryllium salts between beryllium electrodes, in divided cells, gives an unstable univalent Be^+ ion, and the hydrated bivalent Be^{2+} ion is present in aqueous solutions of salts such as the nitrate and sulphate. The Be^{2+} ion is strongly solvated, probably by four water molecules, and it may be extensively hydrolysed. Beryllium forms few truly ionic compounds, however; beryllium oxide, for instance, is non-volatile (m.p. 2,570°C), but it is not a purely ionic substance since it takes up the giant molecule (macromolecular) Wurtzite structure, with Be—O bonds of appreciable covalent character (about 40 per cent according to Pauling). Beryllium fluoride, BeF_2, which melts at *ca.* 800°C, no doubt contains highly polar bonds, but the absence of discrete Be^{2+} and F^- ions is illustrated by the poorly-conducting nature of the fluoride in the fused state.

We have already seen (Chapter 9) that the description of covalent linear beryllium compounds involves *sp* hybrid orbitals. Beryllium chloride, $BeCl_2$, for example, which melts at 404°C and sublimes readily, has no dipole moment in solution and, accordingly, it must have a linear structure in this state. The vapour of the chloride is known to be monomeric at 745°C, but vapour density experiments suggest (RAHLFS[9]) the presence of about 20 per cent of the dimer, Be_2Cl_4, at 564°C. Beryllium bromide has a similar linear structure.

The beryllium atom, however, shows a strong tendency to increase its covalency to a maximum of four, and in solid beryllium chloride, for example (RUNDLE[10]), there is a continuous chain structure in which each beryllium atom is surrounded by four chlorine atoms. The bridging between the beryllium atoms is formally represented

$$Be — Cl = 2·02 Å$$

The bond angles cannot be tetrahedral within the four-membered rings, and the experimental values are Cl—Be—Cl $= 98·2°$ and Be—Cl—Be $= 81·8°$.

A covalency of four is also achieved in co-ordination compounds of the halides with ethers, e.g. $BeCl_2 . 2OEt_2$. The beryllium atom in this compound can again be considered to form two covalent and two co-ordinate links, giving the structure

Here, the initial $BeCl_2$ molecule can be thought of as sp hybridized, but when the ether molecule approaches, the hybridization changes to sp^3, two orbitals being vacant and available to receive the lone-pair electrons of the oxygen atoms.

The structure of the acetylacetone compound of beryllium has been determined by x-ray methods (AMIRTHALINGHAM et al.[11]) and shown to be essentially tetrahedral; the acetylacetone groups form two planes at right angles to one another

The beryllium atom also achieves a covalency of 4 in anionic complex ions such as $[BeF_4]^{2-}$. The Be—F bonds in this ion are often described as ionic, the structure being written

$$\begin{bmatrix} F^- & & F^- \\ & Be^{2+} & \\ F^- & & F^- \end{bmatrix}^{2-}$$

but they are best interpreted as covalent bonds, the beryllium atom making use of sp^3 hybrid orbitals. There are two possible ways of writing down the configuration of beryllium. The first approach is to consider the complex ion to be made up of a Be^{2+} and four F^- ions; the bonds are formed by the donation of the electrons of a lone pair from each fluoride ion into vacant sp^3 hybrid orbitals of the Be^{2+} ion. The Be^{2+} orbitals may be denoted:

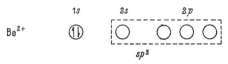

(We shall use this pictorial representation in future; a circle represents the appropriate orbital, capable of accommodating two electrons.) The formation of the $[BeF_4]^{2-}$ ion can be written as follows:

after allocating the requisite formal charges. *Figure 11.1* illustrates the formation of a Be—F bond by the overlap of the doubly-filled

Figure 11.1. Formation of a Be—F bond by the overlap of a vacant sp^3 orbital of Be^{2+} with a doubly-filled p orbital of F^-. (In this, and in all other orbital diagrams, we shall represent a vacant orbital by a dotted outline, a singly-filled orbital by a full line, and a doubly-filled orbital by a full line plus slanting shading. The overlap of orbitals is shown by horizontal shading.)

$2p$ orbital of fluorine with a vacant sp^3 orbital of beryllium. The distorted charge cloud is a convenient way of representing the partial ionic character of the Be—F bond, with the more electronegative fluorine atom having the larger proportion of the electron pair. The alternative way of describing the $[BeF_4]^{2-}$ configuration is to place the double negative charge on the Be atom, so that it has then four singly-occupied sp^3 orbitals which overlap with singly-filled p orbitals of the fluorine atoms. This is shown in *Figure 11.2*.

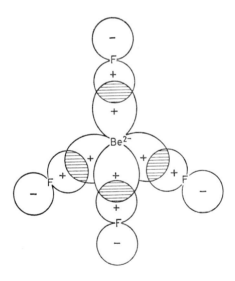

Both of these methods describe 'formal' representations of the bonding process. Of the two, the first is perhaps to be preferred, because it can be applied more simply to complex compounds such as the cobaltammines $[Co(NH_3)_6]^{3+}$, where we picture the bonding as donation of the lone pairs of the ammonia molecules into vacant

Figure 11.2. Formation of the four Be–F bonds in $[BeF_4]^{2-}$, showing overlap of singly-filled sp^3 orbitals of Be^{2-} with singly-filled p orbitals of F

d^2sp^3 orbitals of the cobaltic ion. These complex compounds will be discussed in the next chapter.

The hydrated Be^{2+} ion can be represented in a similar manner the co-ordinate bonds being formed by the donation of the electrons of the oxygen lone pairs into vacant sp^3 beryllium orbitals. It was

$$\left[\begin{array}{c} H_2O \diagdown \quad \diagup OH_2 \\ \quad Be^{2+} \\ H_2O \diagup \quad \diagdown OH_2 \end{array}\right]^{2+}$$

noted that the beryllium ion was the most strongly-hydrated bivalent cation, and this can be partly accounted for by its formation of co-ordinate links with the water molecules. Ions such as Ba^{2+} do not form such bonds; the solvation forces are ion–dipole (*see* Chapter 12) and electrostatic in origin.

BORON AND THE GROUP IIIB ELEMENTS

Boron forms only covalent bonds, but as we proceed down the group the ionic characteristics become more pronounced. Aluminium, for instance, closely resembles beryllium and forms mainly covalent bonds, the ionic character of which is considerably greater than that of the corresponding bonds formed by boron.

We have seen that the neutral boron atom forms trigonal-planar bonds when linked to three similar groups. Thus, all the simple monomeric compounds BX_3 (X=F, Cl, Br, CH_3, OCH_3) have been shown by electron diffraction experiments to be planar with angles of 120°, and, as we saw in Chapter 10, orthoboric acid, $B(OH)_3$, also has a planar arrangement of OH groups about the boron atom, with hydrogen bonding between the $B(OH)_3$ units producing a layer structure. When the groups bonded to boron are dissimilar the angles may be modified slightly, so that with CH_3BF_2 and $C_6H_5BCl_2$ for instance, the XBX angle is 118°. The planar symmetry is also found for the cyclic boron esters

$$\begin{array}{ccc} CH_3 & O & CH_3 \\ \diagdown \diagup & \diagdown \diagup \\ B & & B \\ | & & | \\ O & & O \\ \diagdown & & \diagup \\ & B & \\ & | & \\ & CH_3 & \end{array}$$

although the OBO and BOB angles are appreciably different; we should expect this departure from 120° angles, because the oxygen atom has two lone pairs of electrons which reduce the BOB angle to 112°. Metaboric acid (PETERS and MILBERG[12]) and a number of borates also have a cyclic B_3O_3 skeleton (*e.g.* sodium and potassium metaborates), although a simple planar $[BO_3]^{2-}$ ion is present in cobalt and magnesium borates.

The boron atom also forms planar bonds in the diboron tetrahalides, B_2X_4, (X=F and Cl). These halides are particularly interesting, because although x-ray experiments show that both have the completely planar structures (A) in the solid state, spectroscopic investigations show that the chloride takes up a staggered configuration (B) in the vapour (MANN and FANO[13], HEDBERG and RYAN[14]). It is probable that although the staggered structure

(A)   (B)

represents the lowest energy configuration, in the solid state the planar arrangement may be imposed by crystal lattice forces. The B—F (1·32 Å) and B—Cl (1·74 Å) bond distances are almost the same as those of the simple halides (1·30 and 1·75 Å, respectively).

Boron forms two particularly interesting compounds with nitrogen. The nitride, BN, which is formed as a white crystalline solid by the direct action of nitrogen on boron at white heat, is a giant molecule with a graphite-like structure in which the boron and nitrogen atoms alternate in the rings giving the structure

with B—N bond lengths of 1·45 Å. The basic skeleton bonds are formed by sp^2 hybrid orbitals of the boron and nitrogen atoms; the remaining electrons (represented as forming double bonds) exist in delocalized π orbitals extending above and below the whole plane. The layers are probably arranged so that the B atoms in one layer are immediately over the N atoms in an adjacent layer. A single ring system of boron and nitrogen is found in borazole, $B_3N_3H_6$ (the 'inorganic benzene'), which is formed by the interaction of

B_2H_6 and NH_3 at 200°C.   The structure is:

in which the B—N lengths are 1·44 Å.

The structures of a number of substituted compounds (*e.g.* with N—CH$_3$ and N—Cl bonds) have also proved to be planar with similar B—N distances.   As we might expect, the length of the B—N bond in all these compounds is intermediate between the calculated double (1·36 Å) and single (1·54 Å) B—N lengths.

Just as beryllium uses sp^3 hybrid orbitals to achieve a covalency maximum of four in complex formation, so does boron in such ions as $[BF_4]^-$ and $[BH_4]^-$.   The tetrahedral arrangement is also found in simple co-ordination compounds, and boron trifluoride, for example, accepts electrons from many donors (*e.g.* NH_3, NH_2CH_3, $N(CH_3)_3$, $NCCH_3$, H_2O and $O(CH_3)_2$) to give 1 : 1 compounds in which the FBF bond angles range from 107° to 114°.   It is interesting to note that although BH$_3$ does not exist as a monomer, it can be stabilized by co-ordination with such donors as $N(CH_3)_3$, $P(NH_2)_3$ (NORDMAN[15])   and CO, with the formation of tetrahedral co-ordination compounds.   A cubic form of boron nitride has recently been made (BUNDY and WENTORF[16]) by a reaction at very high temperature and pressure.   It is an extremely hard substance with a giant molecule diamond-type structure in which the boron uses tetrahedrally arranged orbitals.

Aluminium, unlike boron, does not form simple trigonal-planar compounds;   the halides (chloride, bromide, and iodide) are dimeric, each aluminium atom being surrounded by four halogen atoms

Electron diffraction and x-ray studies show that both in the vapour and the solid state the four terminal halogen atoms and the two aluminium atoms are coplanar, with the bridging halogens above and below the plane. Recent nuclear magnetic resonance experiments (GROENEWEGE et al.[17]) have confirmed the presence of similar chlorine—chlorine bridging in $((CH_3)_2ClAl)_2$ and $(CH_3Cl_2Al)_2$. A tetrahedral distribution of chlorine atoms around the aluminium is also found in the $[AlCl_4]^-$ ion. Gallium trichloride (WALLWORK and WORRALL[18]) and tribromide, and indium trichloride, tribromide and tri-iodide (FORRESTER et al.[19]) are also dimeric, with bridged structures analogous to that shown for the aluminium halides, but gallium tri-iodide is reported to be monomeric and planar.

A co-ordination number of four may also be achieved by aluminium and the heavier elements by co-ordination of ligands other than halogen. Thus a recent infra-red study (GREENWOOD et al.[20]) has shown GaH_3,PMe_3 to be monomeric in the vapour phase. The adduct formed by aluminium hydride with trimethylamine, $AlH_3,2NMe_3$ (FRASER et al.[21], HEITSCH et al.[22]), has a trigonal-bipyramidal configuration with the axial positions occupied by amine groups. In $AlH_3,(NMe_2CH_2CH_2NMe_2)$ (PALENIK[23]), aluminium has a similar configuration, with the nitrogen atoms of the amine co-ordinating to different aluminium atoms to give a chain structure. This increase in co-ordination number to values greater than four is not unexpected, since the heavier elements are larger than boron and can make use of d orbitals for additional bonding.

The controversy over the structure of the gallium dihalides has recently been resolved by Raman spectra (WOODWARD et al.[24]) and crystal structure (GARTON and POWELL[25]) studies, which have shown the dichlorides and dibromides to have structures $Ga^+[GaX_4]^-$, with a tetrahedral anion.

CARBON AND THE GROUP IVB ELEMENTS

Carbon forms covalent compounds, except for such molecules as the sodium alkyls, which have a Na^+R^- structure, and the carbides formed by metals in periodic groups I, II, and III, which contain ionic carbide groups. (It should, however, be pointed out that the bonds in these carbides have appreciable covalent character.) This tendency for covalent bonding is also found in the quadrivalent compounds of the other Group IV elements, silicon, germanium, tin, and lead, but the ionic character of the bond increases as we go down the group. There is also an increasing tendency to form bivalent compounds as we go from carbon to lead, two of the four valency electrons exhibiting a decreasing reactivity; they form the so-called 'inert pair'.

In Chapter 9, we saw that quadrivalent carbon may use any one of three possible modes of hybridization, sp, sp^2, and sp^3, exemplified by C_2H_2, C_2H_4, and CH_4; the simple inorganic carbon compounds can be described in a similar way. *Table 11.6* summarizes the position.

Table 11.6. Structure of Carbon Compounds

Hybridization	Shape	σ bonds	Number of lone pairs	π bonds	Examples
sp^3	Tetrahedral	4	0	0	CCl_4
sp^2	Planar	3	0	1	$COCl_2$
sp	Linear	2	0	2	HCN; CO_2

Carbon itself exists in two main modifications, diamond and graphite. In the diamond structure, the carbon atoms use sp^3 orbitals and form a macromolecular tetrahedral structure, with C—C distances of 1·54 Å (as in ethane). With graphite, however, the carbon atoms are arranged in planes, the C—C distances being

Diamond   Graphite

1·42 Å, and the distance between the planes 3·35 Å. It is evident from the physical properties of graphite that layers of carbon atoms are held together by only weak van der Waals forces. Each layer of carbon atoms resembles a vast collection of fused benzene rings, so that the carbon atoms can be considered to be sp^2 hybridized, leaving p orbitals (one per carbon atom) sticking out above and below the plane of the carbon atoms. These p orbitals will produce a double streamer above and below the whole plane. The

C—C length of 1·42 Å is rather longer than the aromatic C—C bond length of 1·39 Å.

The diamond-type of structure is also found in elemental silicon and germanium, and with the low-temperature modification of tin called 'grey' tin. Lead, however, is more metallic and has a cubic structure.

Carbon forms tetrahedral structures in all its halides, CX_4, and in all the saturated hydrocarbons, C_nH_{2n+2}. We have seen, however, that carbon may form only three σ bonds, and give a planar structure, in compounds such as phosgene, $COCl_2$. A similar structure is present in a number of analogous compounds in which the chlorine atoms can be replaced by other univalent atoms or groups, and the oxygen atom by sulphur. The known parameters for a number of these compounds are given in Table 11.7.

Table 11.7. Trigonal-planar Carbon Compounds

Compound (COXX')	XCX'(°)	C=O (Å)	C—X (Å)	References
$COBr_2$	110	1·13	2·05	—
$COCl_2$	111	1·17	1·75	—
CH_3COCl	112·7	1·19	1·79 (C—Cl)	26
COF_2	108	1·17	1·31	27
HCOF	109·9°	1·18	1·31 (C—F)	28

It should be noted that, although the molecules are planar, the bond angles are distorted from the 120° value predicted for sp^2 hybrid orbitals. This distortion can be attributed to the C—O double bond, the electrons in which take up more room than those in the C—X bonds, thus forcing the C—X bonds closer together.

The carbonate ion, CO_3^{2-}, is another planar species, which may be written formally in one of the valence-bond structures as

$$\begin{array}{c} O^- \\ \diagdown \\ \quad C{=}O \\ \diagup \\ O^- \end{array}$$

although all three bonds are in fact equivalent   Because the π bond is delocalized, the effective π-bond order is only 0·33, and the bond length ($\sim$ 1·29 Å) is accordingly much greater than in $COCl_2$. A simple description of the carbonate ion involves carbon sp^2

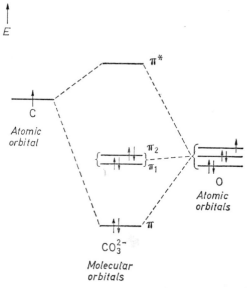

Figure 11.3. Molecular orbital energy diagram for $[CO_3]^{2-}$
(All atoms use $2p_z$ a.o.'s)

hybrids for the σ bonding, together with the π molecular-orbital scheme shown in *Figure 11.3*. This shows that four of the π electrons are in non-bonding orbitals (π_1 and π_2) and the other two in the bonding π orbital. The latter, which is given by the combination

$$\psi\pi = \psi C_{2p_z} + \psi Oa_{2p_z} + \psi Ob_{2p_z} + \psi Oc_{2p_z}$$

is illustrated in *Figure 11.4*.

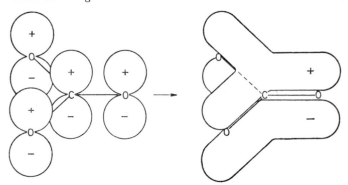

Figure 11.4. Bonding π orbital in $[CO_3]^{2-}$

168

The carbon atom also uses sp hybrid orbitals to form two linear bonds, thus leaving two p electrons for π bonding. Because of the alternative methods of using these π electrons, the carbon atom may either form two double bonds (A), or form one single and one triple bond (B).

$$A \quad X{=}C{=}X$$
$$B \quad Y{-}C{\equiv}Z$$

The structures of quite a number of compounds of type A have been studied, *e.g.* CO_2, COS, COSe, CONH, CS_2, CSSe, CSTe, and CSNH, and accurate structure determinations show that the $C{=}O$ and $C{=}S$ lengths remain constant throughout this range of compounds, being 1·16 Å and 1·56 Å, respectively. The zero dipole moment found for CO_2 and CS_2 is in agreement with the linear bonding arrangement. The simple double-bonded formulation is inadequate, however, because the measured bond lengths are intermediate between those calculated for double and triple bonds (1·22 Å and 1·10 Å, respectively, for C—O). In a valence-bond treatment, therefore, we consider contributions from other possible structures:

$$X^+{\equiv}C{-}X^- \quad \text{II}$$
$$X^-{-}C{\equiv}X^+ \quad \text{III}$$
$$X^-{-}C^+{=}X \quad \text{IV}$$
$$X{=}C^+{-}X^- \quad \text{V}$$

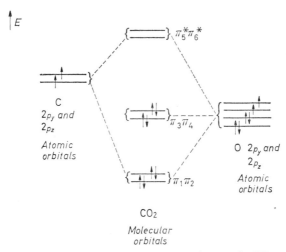

CO_2

Molecular orbitals

Figure 11.5. π molecular orbital energy diagram for CO_2

although the last two structures make very small contributions because they correspond to higher energy states.

The carbon dioxide molecule may be described in terms of σ bonding based on sp hybrids, plus delocalized π bonding based on the scheme shown in *Figure 11.5*. Both the bonding (π_1 and π_2) and non-bonding (π_3 and π_4) orbitals are fully occupied. In

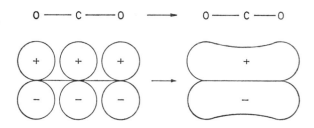

Figure 11.6. Bonding π molecular orbital for CO_2, showing asymmetry

Figure 11.6, one of the bonding π orbitals is shown as distorted so as to give a greater charge density over the oxygen atoms; this is a reflection of the oxygen $2p$ orbitals being of lower energy than their carbon analogues and contributing more to the molecular wave function.

The isocyanate ion, NCO^-, which is isoelectronic with carbon dioxide, is also linear, with bond lengths C—O and C—N of 1·18 and 1·20 Å respectively (BRITTON and DUNITZ[29]).

The sub-oxides of carbon have similar structures. Thus electron diffraction studies of C_3O_2 (LIVINGSTON and RAO[30]) give C—C = 1·28 Å and C—O = 1·16 Å, which correspond to a molecular structure that is most easily described as a hybrid of the forms:

$$O=C=C=C=O \qquad I$$
$$O^+ \equiv C—C \equiv C—O^- \qquad II$$
$$O^-—C \equiv C—C \equiv O^+ \qquad III$$

An orbital diagram similar to that in *Figure 11.5* could be drawn up. An oxide, C_5O_2, presumably containing a chain of five carbon atoms, has been reported, although its existence has not been firmly established. It has been suggested that oxides of formula C_nO_2 can exist only when n is odd, because the type of resonance suggested for the above structures of C_3O_2 could not be present if an even number of carbon atoms were present in the chain. Oxides with the formulae C_2O_2, C_4O_2, etc., have not so far been discovered.

Linear structures of type B, with a triple bond, are found in hydrogen cyanide, methyl cyanide, and the bromo-, chloro- and fluoro-derivatives $(X—C\equiv N)$; in these compounds the carbon-nitrogen bond length remains constant at $1·16$ Å, a value which would be expected for a triple bond.

A study (TYLER[31]) of the microwave spectrum of HCP (the phosphorus analogue of HCN) has confirmed the expected linear configuration, with $H—C = 1·07$ Å and $C—P = 1·54$ Å, and shown that the molecule has quite a small dipole moment ($\mu = 0·39$ D) compared to HCN ($\mu = 2·9$ D). The most obvious reason for this smaller moment is that phosphorus is much less electronegative, than nitrogen, although a contributory factor could be that while the nitrogen lone pair is in an sp hybrid orbital (along the bond axis), the phosphorus lone pair will probably be in an orbital with much less p character.

The cyanogen molecule, $(CN)_2$ (PARKES and HUGHES[32]), also has a linear arrangement, with bond lengths $C—C = 1·37$ Å and $C—N = 1·13$ Å, but the C—C length is too short for a single bond and there is evidently some delocalization of the C—N π electrons.

The structure of the HCN tetramer has been determined (PENFOLD and LIPSCOMB[33]), and this is interesting in that there are two different structural types of carbon atom, one with a trigonal-planar and the other with a linear configuration:

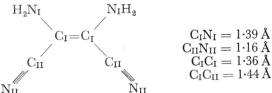

$$C_IN_I = 1·39 \text{ Å}$$
$$C_{II}N_{II} = 1·16 \text{ Å}$$
$$C_IC_I = 1·36 \text{ Å}$$
$$C_IC_{II} = 1·44 \text{ Å}$$

The $C_{II}—N_{II}$ bonds are evidently triple, but the others ($C_I—N_I$) appear to be part way between a single and a double bond; the $C_I—C_I$ bond is rather long for a true double bond and the $C_I—C_{II}$ bonds are short for single bonds, so it would seem that there must be a certain amount of delocalized π bonding over the entire C_4N_4 skeleton.

Unlike carbon, silicon does not form trigonal-planar or linear structures but is always bonded to at least four other atoms, giving configurations that are basically tetrahedral. Thus the hydrides, alkyls and halides are all essentially tetrahedral. The dioxide and disulphide are not monomeric like their carbon analogues, but give macromolecular systems with an approximately tetrahedral

171

arrangement of oxygen and sulphur atoms about the silicon atoms. Various structures are known for silica (quartz, cristobalite, tridymite), which differ only in the way in which the tetrahedral SiO_4 units are linked to each other by shared oxygen atoms. The silicate ion, $[SiO_3]^{2-}$, in sodium silicate also forms a linked tetrahedral system

and the many apparently complex silicate structures can be discussed in terms of chains, sheets and three-dimensional structures of linked SiO_4 tetrahedral units.

As we might expect, the bond angles in such structures are not all the same, being greatest between the external Si—Ō bonds (119°) where repulsion is greatest, and least between the linked Si—O bonds (101°). The disulphide, SiS_2, a solid crystallizing in long silky needles, consists of infinite chains of the type

The silicon atom is capable of achieving an octahedral configuration in complexes such as $[SiF_6]^{2-}$ and $SiCl_4,2C_5H_5N$, because the vacant $3d$ orbitals are energetically available to accept electrons from suitable donors (see Chapter 12).

In the quadrivalent state, both tin and lead form tetrahedral structures in the tetrachlorides and tetra-alkyls, and attain an octahedral configuration in a range of co-ordination compounds (*e.g.* $[MCl_6]^{2-}$).

The divalent state is important only for tin and lead. The tin compounds $SnCl_2,2H_2O$ and K_2SnCl_4,H_2O have been examined (*cf.* RUNDLE and OLSON[34]) and shown to contain pyramidal $SnCl_2,H_2O$ and $[SnCl_3]^-$ species, respectively. The bond angles are slightly less than 90°, and the structures can best be regarded as involving pure p overlap with the lone (inert) pair electrons in the $5s$ orbital.

NITROGEN AND THE GROUP VB ELEMENTS
Nitrogen

Nitrogen, with the configuration $1s^2 2s^2 2p^3$, has three unpaired electrons, and it is accordingly tervalent in its simple compounds; the next suitable lowest energy orbital, the $3d$, is too far away to allow the promotion of one of the $2s$ electrons, so that the neutral nitrogen atom never exceeds a covalency of three. We saw at the beginning of this chapter that the distribution of these three bonds can be considered tetrahedral, with a lone pair of electrons occupying one position. A positively-charged nitrogen atom, on the other hand, is isoelectronic with a neutral carbon atom and can have four unpaired electrons, while a negatively-charged nitrogen atom is isoelectronic with a neutral oxygen atom, and we get two bonds and two lone pairs of electrons. There are also other trigonal-planar and linear structures which arise from bonds formed by sp^2 and sp hybrid orbitals, respectively. *Table 11.8* summarizes the structural chemistry of nitrogen.

(*a*) sp^3 *hybridization*—The tetrahedral distribution of bonds about a positively-charged nitrogen atom has been established for the ammonium ion $[NH_4]^+$, and the tetra-alkylammonium ions $[NR_4]^+$. Trimethylamine oxide, in which nitrogen is formally positive (*i.e.* $Me_3N^+—O^-$), is also tetrahedral (CARON *et al.*[35]) with $N—O = 1.40$ Å.

The neutral nitrogen atom forms three N—H bonds in the ammonia molecule, with HNH angles of 107·3°. The deviation of the bond angles from tetrahedral (109·5°) arises from the presence of the lone pair of electrons, because these electrons occupy more volume than do those in the nitrogen-hydrogen bonds. MELLISH and LINNETT[36] have pointed out that the bonding pairs and the lone pair will tend to arrange themselves so as to give maximum mutual separation, thus giving HNH angles rather less than the expected tetrahedral value. The corresponding NF_3 molecule has FNF angles of 102·1°. This is a bigger deviation from the tetrahedral angle than that found in ammonia, but expected, because the fluorine atoms are more electronegative than the hydrogen atoms. The electrons in the N—F bonds will be pulled further from the nitrogen atom, and restricted more closely to the line of the nuclei (giving a 'thinner' molecular orbital, which occupies less volume and leaves more of the space around the nitrogen atom available for the lone pair). The N—F bonds in NF_3 accordingly close up together (as compared with the N—H bonds in NH_3). The wave functions which describe the bonds

Table 11.8. Structures of Nitrogen Compounds

Hybridiza-tion	Shape	Number of σ bonds	Number of lone pairs	Number of π bonds	State of N	Example
sp^3	Tetrahedral	4	0	0	N^+	$[NH_4]^+$
	Trigonal-pyramidal	3	1	0	N	NH_3
	Angular	2	2	0	N^-	$[NH_2]^-$
sp^2	Trigonal-planar	3	0	1	N^+	$\begin{array}{c} O^- \\ {\diagdown} \\ O^- \end{array} N^+{=}O$
	Angular	2	1	1	N	$\begin{array}{c} Cl \\ {\diagdown} \\ \quad N{=}O \end{array}$
sp	Linear	2	0	2	N^+	$\left.\begin{array}{c} \\ \\ \end{array}\right\} N^-{=}N^+{=}O$
		1	2	1	N^-	

in ammonia and nitrogen trifluoride will not represent simple sp^3 orbitals; the N—F orbitals are said to have more p character than the N—H orbitals. It should be realized, however, that the exact description of the orbitals used (*i.e.* amount of p character) is a consequence of the arrangement of electrons for minimum repulsion.

The dipole moments of NH_3 (1·5D) and NF_3 (0·2D) can be explained (COULSON[37], BURNETTE and COULSON[38]) by attributing a dipole to the strongly directional lone pair (*cf. Figure 11.7*). Thus

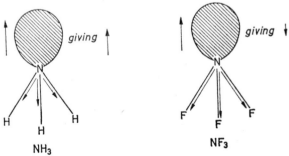

NH₃   NF₃

Figure 11.7. Dipole moments for NH_3 and NF_3

the resultant moment of the three N—H or N—F bonds is in the opposite direction to that of the lone pair, but whereas the NF_3 bond dipole just about cancels out that of the lone pair, the NH_3 bond dipole is much less, and a significant overall moment remains. (The dipole of the N—H bond is in the direction shown because the two orbitals used—sp^3 for N and s for H—are different in size and overlap much closer to hydrogen than to nitrogen.)

Substituted ammonia molecules such as chloramine, NH_2Cl, and hydroxylamine, NH_2OH, also have pyramidal shapes, the HNH angles being 102° and 107°, respectively; the NHF_2 molecule has an FNF angle of 102·9° (Lide[39]). In dinitrogen tetrafluoride, N_2F_4, the two NF_2 groups are linked through the nitrogens, both halves being pyramidal (Lide and Mann[40]), but for the related compound, $N_2(CF_3)_4$, electron diffraction experiments (Bartell and Higginbotham[41]) indicate a skew structure analogous to that of B_2Cl_4 (see page 143), with almost planar configurations about the nitrogen atoms; this increase in the bond angle to 120° evidently results from repulsions between the bulky CF_3 groups.

(b) sp^2 hybridization—When a nitrogen atom forms three σ and one π bond, the resulting structure is planar. Dinitrosomethane (Germain *et al.*[42]) illustrates this

$$N{=}N \quad 1{\cdot}31 \text{ Å}$$
$$N{-}O \quad 1{\cdot}31 \text{ Å}$$

Another illustration is the nitrate ion $[NO_3]^-$, in which all the N—O bond lengths are 1·26 Å (Truter *et al.*[43]), a value which is midway between the values expected for double and single N—O bonds. The valence-bond approach enables us to get a picture of the ion in terms of the three forms:

I II III

The π bond is distributed over the three N—O bonds, thus accounting for the observed bond lengths. The molecular-orbital approach gives us a deeper insight, however, and here we use the scheme shown for the carbonate ion in *Figure 11.3* and obtain precisely the same configuration with two bonding and four non-bonding π electrons.

The structure of the parent acid, HNO_3, has been determined by several methods (STERN et al.[44]), the parameters obtained from the x-ray study being

HO
$\diagdown \sim 1{\cdot}30$ Å
$\diagdown \qquad \sim 1{\cdot}24$ Å
N——O
$\diagup 134°$
O

Microwave (Cox and RIVEROS[45]) and electron diffraction studies agree with this general picture but assign values of $1{\cdot}41$ and $1{\cdot}21$ Å to the N—O bonds; the proton is clearly associated with the long N—O bond. The other two N—O bonds have considerable π character (as for the —NO_2 group discussed in Chapter 9), and repulsion between them results in the ONO angle being appreciably greater than $120°$.

Analogous structures are formed by molecules of the general type NO_2X, where X = Cl, F, Me, NH_2, OCl and OF, the known parameters being summarized in *Table 11.9*.

Table 11.9. Parameters for Molecules of the Type XNO_2

Compound	ONO (°)	N—O (Å)	N—X (Å)
NO_2Cl	130·1	1·20	1·84
NO_2F	130	1·21	1·40
$MeNO_2$	130	1·20	—
$NO_2(NH_2)$	130·1	1·18	1·40
$NO_2(OF)$	125	—	—

There are quite a number of angular structures, of the general type XNO, XNCO and XNCS, in which one of the trigonal bonds is replaced by a lone pair; the known parameters are shown in *Table 11.10*.

X
$\diagdown$
$\diagdown$
$N{=}O$
$\diagup$

X
$\diagdown$
$\diagdown$
$N{=}C{=}O$
$\diagup$

Table 11.10. Parameters for XNO *and* XNCO *Molecules*

Compound	XNO (°) (or XNC)	X—N (Å)	N—O or N—C (Å)
BrNO	114	2·14	1·15
ClNO	114	1·95	1·14
FNO	110	1·52	1·17
HONO	118	—	—
HNCO	128·1	0·99	1·21
CH₃NCO	125	—	1·19
HNCS	130·3	1·01	1·22
CH₃NCS	142	1·47	1·22

The nitrite ion, $[NO_2]^-$, is also angular ($<ONO = 114\cdot9°$, $N—O = 1\cdot24$ Å; KAY and FRASER[46]) and may be regarded as a nitro group with two bonding and two non-bonding π electrons.

The microwave spectrum (KUCZKOWSKI and WILSON[47]) of cis-N_2F_2 shows the molecule to have a similar structure

N—F 1·38 Å
N—N 1·21 Å

Nitrogen dioxide, NO_2, is an especially interesting molecule with an angular structure, because it contains an unpaired electron which has been shown by electron-spin resonance experiments to be in a σ rather than a π orbital. We can, therefore, formally consider the molecule as an $-NO_2$ grouping but with the odd electron in the sp^2 hybrid orbital, *i.e.*

The ONO angle is 134° and the N—O bond length 1·20 Å. In addition to the σ bonding based on sp^2 hybrid orbitals, there will be four electrons in the π system, two bonding and two non-bonding (*cf.* Chapter 9).

The pairing up of two nitrogen dioxide molecules leads naturally to dinitrogen tetroxide, which has a planar skeleton (SMITH and HEDBERG[48]) with the dimensions

177

$$
\begin{array}{ccc}
& 1\cdot 18\ \text{Å} & \\
\text{O} & & (\ \text{O} \\
\diagdown & & \diagup \\
& \text{N}\text{------}\text{N}\ ) & 133\cdot 7° \\
\diagup & 1\cdot 75\ \text{Å} & \diagdown \\
\text{O} & & \text{O}
\end{array}
$$

We can formally regard the molecule as consisting of two NO_2 groups held together by a single σ bond, the π bonding being limited to the individual ONO systems. This is undoubtedly oversimplified because it does not explain why the molecule is planar rather than eclipsed, and why there is a large energy barrier to rotation about the N—N bond. We could account for these points by supposing that the π bonding was delocalized over the entire skeleton, but then we have to explain why the N—N bond is so long when it contains considerable π character. Some lengthening would of course be expected, because the two nitrogen atoms have formal positive charges. Some workers (COULSON and DUCHESNE[49]) have proposed a linking of NO_2 groups by π bonding only, no σ bonds being involved, but others (GREEN and LINNETT[50]) prefer the σ model. The bonding is obviously still by no means settled, and interested readers are referred to a recent review of the problem by BENT[51].

A thorough survey of the available spectroscopic information (MASON[52]) for N_2O_3 indicates that the molecule has a nitro-nitroso structure, with one nitrogen atom planar and the other angular

$$
\begin{array}{ccc}
\text{O} & & \text{O} \\
\diagdown & & \diagup \\
& \text{N—N} & \\
\diagup & & \diagdown \\
\text{O} & &
\end{array}
$$

(c) sp hybridization—We now discuss linear structures, in which the nitrogen atom forms two π bonds and two σ bonds, as in the azide ion $[N_3]^-$, the nitronium ion $[NO_2]^+$, and nitrous oxide, N_2O.

In the azide ion, the two N—N distances are the same, being $1\cdot 12$ Å in $Sr(N_3)_2$ for instance. The parent acid, hydrazoic acid, HN_3, is covalent; it has a linear N—N—N skeleton, but the two N—N bonds are no longer equivalent (WINNEWISSER and COOK[53])

$$
\begin{array}{cccc}
& 1\cdot 24\ \text{Å} & 1\cdot 13\ \text{Å} & \\
& \text{N}\text{------}(\text{N}\text{------}(\text{N} & & \\
0\cdot 98\ \text{Å} \sim \diagup & 114\cdot 1° & & \\
\text{H} & & &
\end{array}
$$

Methyl azide has a similar structure with N—N distances of 1·24 Å and 1·12 Å (LIVINGSTON and RAO[54]). The usual interpretation of these structures is a valence-bond one, the main contributing structures being

$$H—N=N^+=N^- \qquad I$$

and

$$H—N^-—N^+ \equiv N \qquad II$$

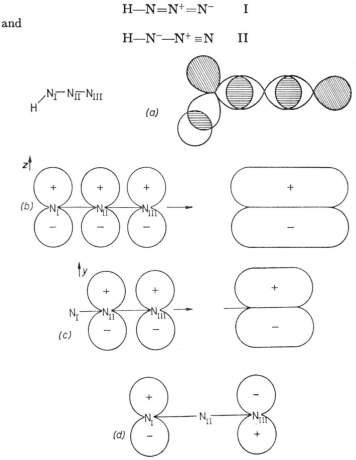

Figure 11.8. Orbitals for HN_3: (a) σ orbitals; (b) π_1 orbital; (c) π_2 orbital; (d) π_3 orbital

The σ bond system for HN_3, which is shown in *Figure 11.8 (a)*, proposes the use of sp^2 hybrid orbitals by the nitrogen atom (N_I) attached to hydrogen, and sp hybrids by the other two nitrogen atoms (N_{II} and N_{III}). The first three π molecular orbitals are

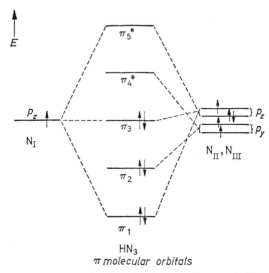

HN$_3$
π molecular orbitals

Figure 11.9. π molecular orbital energy diagram for HN$_3$

illustrated in *Figure 11.8 (b)–(d)*, the π electron energy levels are shown in *Figure 11.9*. It can be seen that there is one delocalized π orbital (π_1) based on the three nitrogen p_z atomic orbitals, one localized π orbital (π_2) obtained by the combination of the N$_{II}$ and N$_{III}$ p_y orbitals, and one non-bonding π orbital based on N$_I$ and N$_{III}$.

Nitrous oxide, which is also linear, has been shown by molecular spectroscopy to have the nuclei arranged N—N—O, rather than N—O—N, the bond lengths being

$$N—N = 1{\cdot}13 \text{ Å}$$

and

$$N—O = 1{\cdot}19 \text{ Å}.$$

The valence-bond description of this molecule is a combination of the structures

$$N^-{=}N^+{=}O \qquad I$$
$$N \equiv N^+{—}O^- \qquad II$$

We can assume that the nitrogen and oxygen atoms use sp hybrid orbitals, and that an orbital interpretation can be made by superposing the orbital forms corresponding to structures I and II. There will be one localized π orbital between the two nitrogen atoms, and a delocalized π orbital extending over all three atoms.

The two remaining electrons will occupy a non-bonding orbital, similar to that described for HN_3. There are, in addition, one lone pair of electrons on the terminal nitrogen atom (sp orbital) and two lone pairs of electrons on the oxygen atom (one sp and one p orbital).

Solid dinitrogen pentoxide has been found to be ionic, with $[NO_2]^+$ and $[NO_3]^-$ ions. The nitronium ion is linear, $O=N^+=O$, as would be expected, with N—O bond lengths of 1·15 Å (TRUTER et al.[43]).

Finally, we can refer again to the nitric oxide molecule (cf. Chapter 7, page 87) which has an N—O length of 1·10 Å. We saw that an adequate molecular-orbital description of the molecule placed this unpaired electron in an antibonding molecular orbital. The alternative description of NO as a hybrid of forms I and II is less satisfactory; it gives a three-electron bond between the nitro-

$$\dot{N}=O \qquad I$$
$$N^-=\dot{O}^+ \qquad II$$

gen and oxygen atoms. Nitric oxide is essentially a free radical, but the unpaired electron is distributed over the whole molecule, so that the 'sensitivity' and tendency for pairing of this electron is thus diminished. A dimer forms at low temperatures, but the bonding is weak, and GREEN and LINNETT[55] point out that this must be expected since there is no increase in bond order on dimer formation.

Phosphorus, Arsenic and Antimony

Whereas nitrogen itself is a diatomic gas, phosphorus is a solid at room temperature; P_4 molecules are found in the vapour phase, however, and above 800°C a dissociation into P_2 molecules takes place. Electron diffraction shows that the phosphorus atoms in the P_4 molecule are at the corners of a regular tetrahedron, the P—P bond distances being 2·21 Å and the P—P—P angles being 60°. HART et al.[56] have recently used a molecular-orbital approach and come to the conclusion that the best description of the bonding involves phosphorus pure p orbitals rather than pd^2 hybrid orbitals as proposed earlier by ARNOLD[57]. Arsenic also forms As_4 molecules in the vapour phase, but the position is rather more obscure with antimony and bismuth, although it seems likely that Sb_4 and Bi_2 molecules are present in the vapours of the respective elements.

In many of their compounds, phosphorus, arsenic, antimony and bismuth differ from nitrogen in that they can use d orbitals and increase their covalency to five or six. VAN WAZER[58] has discussed the fundamental aspects of bonding in phosphorus compounds, and from nuclear magnetic resonance experiments he concludes that when phosphorus is attached to three, five or six atoms, the links are essentially single bonds, whereas when four atoms are attached the links have some double-bond character. The essential structural chemistry of the Group VB elements (other than nitrogen) is summarized in *Table 11.11*.

Table 11.11. Structures of Compounds of the VB Elements

Hybridiza-tion	Shape	Number of			Examples
		σ bonds	lone pairs	π bonds	
sp^3	Tetrahedral	4	0	0	$[PH_4]^+$, $[AsPh_4]^+$
	Trigonal-pyramidal	3	1	0	PF_3, $AsCl_3$, $SbCl_3$, $BiBr_3$
	Tetrahedral	4	0	1	$POCl_3$
sp^3d	Trigonal-bipyramidal	5	0	0	PCl_5, AsF_5, $SbCl_5$
sp^3d^2	Octahedral	6	0	0	$[PF_6]^-$, $[AsF_6]^-$, $[SbCl_6]^-$ $[BiCl_6]^-$
	Square-pyramidal	5	1	0	$[SbF_5]^{2-}$

(a) sp^3 hybridization—Positively-charged phosphorus (which is isoelectronic with silicon) and arsenic form simple tetrahedral structures in such ions as $[PH_4]^+$, $[PBr_4]^+$, $[PCl_4]^+$ and $[AsPh_4]^+$.

With the neutral tervalent elements, we have a situation similar to that found for nitrogen in which one tetrahedral position is taken

Table 11.12. XMX Bond Angles in MX_3 Compounds

Compound	P	As	Sb
MBr_3	100	100	96
MCl_3	100	100	95·2
$M(CN)_3$[59a,b]	93	92	—
MF_3	104	102	—
MH_3	93·3	91·8	91·3
MI_3	98	98·5	99·1
MMe_3	99·1	96	—

up by a lone pair of electrons. As the values quoted in *Table 11.12* show, the XMX bond angles in the MX_3 type compounds are appreciably less than the tetrahedral value, and in the hydrides the angles are little more than 90°.

The much smaller bond angle found in the hydrides (compared with NH_3) may possibly be attributed to the increased size of the phosphorus, arsenic and antimony atoms, which permits a closer approach of the bonding pairs before their mutual repulsions become important, and allows the lone-pair electrons to expand. GILLESPIE[60] has suggested that the natural structure for second and subsequent row elements is octahedral, so that the angle between the bonding pairs will approach 90° before repulsions become important. With the halides, repulsion between lone pairs on neighbouring halogen atoms will tend to keep the bond angles in the region of 100°. Similar but less symmetrical pyramidal structures are observed for compounds such as PCl_2F, Me_2PH and $MeAsI_2$.

Quinquevalent phosphorus forms a number of tetrahedral complexes of the type $X_3P=Y$ (where X=alkyl, aryl, Br, Cl and F and Y=O and S) where four electrons are used for σ bonding and one electron for π bonding. Whereas the nitrogen–oxygen bond in

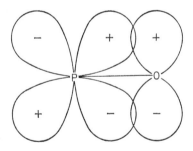

Figure 11.10. d–$p_{\pi\pi}$ bonding

R_3NO compounds has to be written N → O or N⁺—O⁻, because nitrogen cannot form more than four covalent bonds, the PO or PS bonds can be considered essentially double because the $3d$ orbitals of phosphorus are now available for π bonding. This d_π–p_π type bonding is shown in *Figure 11.10*. Such double bonds should be appreciably shorter than single bonds, and this may be illustrated by reference to $(PhO)_3PO$ in which the 'double' bond length is 1·32 Å and the three single bonds are 1·63 Å in length (DAVIES and STANLEY[61]). In all these cases, the XPX angles are less than the tetrahedral value, ranging in fact from 100° to 108°, but this should be expected because the double bond electrons will repel the single bond pairs more than the latter repel each other. Thus with tervalent phosphorus we get three σ bonds and a lone pair, while with quinquevalent phosphorus we get a double bond instead of the lone pair.

The 'tetrahedral' arrangement is found also in a range of oxycompounds of phosphorus, such as P_4O_6 and P_4O_{10}, with O—P—O bond angles of 99° and 101·5°, respectively.

P_4O_6 P_4O_{10}

In P_4O_{10} the terminal and skeleton P—O bonds are 1·40 and 1·60 Å, respectively, in accordance with the double- and single-bond formulation (CRUIKSHANK[62], AKISTIIN et al.[63]).

The structure of phosphorous acid is interesting in that two P—O bonds (1·54 Å) are longer than the third (1·47 Å), values agreeing with the view that the short bond is P=O, and that a hydrogen atom is attached to the other two oxygen atoms

The angle between the P—OH bonds is 102°, and those between the P=O and P—OH bonds are 113° and 116°, respectively, values consistent with a greater repulsion exerted by electrons in the double bond. The individual $HP(OH)_2O$ molecules are linked together by hydrogen bonds. Simple orthophosphate ions, such as $[H_2PO_4]^-$ and $[HPO_4]^{2-}$, have a tetrahedral configuration, as do more condensed phosphate ions such as $[P_2O_7]^{4-}$, $[P_3O_{10}]^{5-}$, and $[P_4O_{12}]^{4-}$. The first two condensed ions form chain structures and the latter has a cyclic arrangement. Thus we can write the structure for the triphosphate ion, although it must be appreciated that this is only one form and that all the terminal bonds are equivalent with some π bond character (P—O $= 1{\cdot}50$ Å); the P—O bonds forming the skeleton are much longer ($1{\cdot}61$ Å and $1{\cdot}68$ Å). Similar values ($1{\cdot}49$ Å and $1{\cdot}61$ Å) are found for the terminal and skeleton P—O bonds in the cyclic $[P_4O_{12}]^{4-}$ ion.

Sulphur compounds of phosphorus can be obtained in which some or all of the oxygen atoms in the P_4O_{10} structure have been replaced by sulphur atoms. Thus $P_4O_6S_4$ and P_4S_{10} are well known; in the first compound, the four sulphur atoms take the place of four of the terminal oxygen atoms. P_4S_6 does not appear to exist, but a range of sulphides (P_4S_n, $n = 3$, 5, and 7) has been characterized, and shows the following structures:

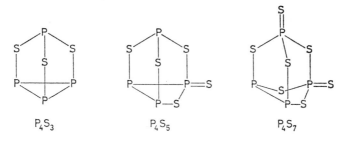

The internal and external P—S bonds have values close to $2{\cdot}09$ Å and $1{\cdot}95$ Å, respectively, in all these compounds, in agreement with our assignment of π bonding to the external bonds. In these sulphides, the S—P—S angles (internal) range from the tetrahedral value in P_4S_{10} down to 99° in P_4S_3. The selenide, P_4Se_3, has been shown to have a structure analogous to that of P_4S_3.

Arsenic also achieves a tetrahedral configuration in the simple arsenate ion $[AsO_4]^{3-}$, while one modification of arsenic(III) oxide (As_4O_6) has a structure similar to that of P_4O_6. An essentially tetrahedral configuration is found in linked polyarsenites such as ($NaAsO_2)x$, in which a lone pair of electrons takes up one bonding position:

185

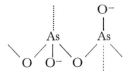

Antimony has a tetrahedral arrangement in the oxide, Sb_4O_6, and the thio-ion $[SbS_4]^{3-}$.

The phosphorus atom also forms four σ bonds in cyclic phosphorus–nitrogen compounds, such as the phosphonitrilic halides $(PNX_2)_n$. *Table 11.13* summarizes the structural data, from which it may be seen that while the two trimers are essentially planar, the tetramers are puckered to varying extents.

Table 11.13. Parameters for Phosphonitrilic Compounds

Compound	P—N (Å)	NPN (°)	PNP (°)	XPX (°)	Reference
$(PNCl_2)_3$	1·60	120·4	120·9	102	1
$(PNCl_2)_4$	1·57	121	131	103	64
$(PNF_2)_3$	1·57	119·4	120	99·1	65a
$(PNF_2)_4$	1·51	122·7	147·2	99·9	66
$(PNMe_2)_4$	1·59	119·8	132	104·1	65b
$(PN(NMe_2)_2)_4$	1·59	121	129	102	67

The configuration about each phosphorus atom may be considered as a distorted tetrahedron. Except in $(PNF_2)_4$, the P—N length is remarkably constant and falls within the range 1·57–1·60 Å, which is very much less than the calculated single bond value of 1·84 Å. The shortening is attributed (CRAIG [68]) to delocalized d_π–p_π bonding arising from the lateral overlap of the d_{xz} orbitals of phosphorus with the p_z orbitals of nitrogen

DEWAR *et al.*[69] have proposed a more restricted π-bonding that embraces only three atoms, two phosphorus and one nitrogen; the 3-centre π orbital uses the same p_z orbital of nitrogen, but a linear combination of the d_{yz} and d_{xz} orbitals for phosphorus (to give better overlap). Dewar points out that this description will apply

186

to non-linear structures, which must be present in the higher members of the series, and is consistent with the similarity of spectra found for all the compounds. CRAIG and MITCHELL[70] have discussed this problem of 'island' and 'cyclic' $d_\pi-p_\pi$ bonding for the more general case of compounds consisting of rings of alternating first- and second-row elements, and interested readers are referred to this paper.

(b) sp³d hybridization—If the quinquevalent elements form only σ bonds, their compounds (MX_5, etc.) would be expected to have trigonal-pyramidal configurations. Such bonding arrangements have been found for a number of the pentahalides in the gas phase, *e.g.* PF_5, PCl_5, PBr_5, AsF_5 and $SbCl_5$. The alkyl derivatives of the antimony halides (R_3SbX_2; $X=Cl$, Br and I) appear to have trigonal-bipyramidal structures with the alkyl groups in the equatorial positions. Over the past few years, several sets of workers (HOLMES *et al.*[71, 72], GRIFFITHS[73] and MUETTERTIES *et al.*[74]) have made spectroscopic (infra-red, Raman and nuclear magnetic resonance) studies on various substituted pentahalides of phosphorus, arsenic, antimony and bismuth, and they conclude that the molecules are invariably trigonal-bipyramidal, with the more electronegative groups occupying the axial positions. Thus the axial positions are occupied by fluorine atoms in PCl_3F_2 and PCl_2F_3.

The pentaphenyls are curious (WHEATLEY[75]) in that, while the phosphorus and arsenic compounds are trigonal-bipyramidal, the antimony compound is described as a square pyramid; the configuration is not a regular one, however, since all the bond lengths differ (ranging from 2·05 to 2·23 Å) and the antimony atom is well above the basal plane. We shall return to this problem of the bonding arrangements in such compounds in the next chapter.

The pentachloride and pentabromide of phosphorus have ionic structures in the solid state, however, being $[PCl_4]^+$ $[PCl_6]^-$ and $[PBr_4]^+$ Br^-, respectively. The positive ions are tetrahedral, as we have already seen, and the $[PCl_6]^-$ ion is octahedral, so that the trigonal-bipyramidal structure found in the gas phase has become a mixture of the more symmetrical ions in the solid state.

(c) sp³d² hybridization—All the Group VB elements form hexahalogeno ions $[MX_6]^-$ in which there are six σ bonds octahedrally arranged, and a number of complex ions with this co-ordination is also known; thus antimony(V) chloride forms a compound $SbCl_5 \leftarrow OPCl_3$, in which the oxygen atom donates a lone pair of electrons to antimony and completes the octahedral structure.

In the ion $[SbF_5]^{2-}$, the five fluorine atoms occupy five of the six octahedral positions, with a lone pair of electrons taking up the sixth position; repulsions between the antimony-fluorine bonding pairs and the lone pair result in the antimony atom being somewhat below the plane of the four fluorine atoms (GRDENIĆ and ŠĆAVNI-ČAR [76]). An account of the configurations possible in 5-co-ordinate systems such as this will be given in the next chapter.

OXYGEN AND THE GROUP VIB ELEMENTS

Oxygen

The neutral oxygen atom forms only two covalent bonds, in accordance with its $1s^2 2s^2 2p_x^2 2p_y^1 2p_z^1$ configuration; the promotion of one of the $2p_x$ electrons into an orbital of principal quantum number 3 is not possible in view of the considerable energy which is required. Upon acquiring a single positive charge, the oxygen atom becomes isoelectronic with a neutral nitrogen atom and forms three angular bonds, *e.g.* H_3O^+ (NORDMAN [77]). A negatively-charged oxygen atom, on the other hand, now contains only one unpaired electron and consequently forms only one bond, *e.g.* OH^-.

We have already discussed the structure of the diatomic oxygen molecule in Chapter 7, where we saw that it could best be given the molecular-orbital configuration

$$O_2[KK(z\sigma)^2(y\sigma)^2(x\sigma)^2(w\pi)^4(v\pi)^2]$$

Table 11.14 summarizes the configurations of oxygen compounds.

Table 11.14. Configurations of Oxygen Compounds

		Number of				
Hybridization	*Shape*	σ bonds	*lone pairs*	π bonds	*State of* O	*Example*
sp^3	Trigonal-pyramidal	3	1	0	O^+	$[OH_3]^+$
	Angular	2	2	0	O	OH_2
sp^2	Angular	2	1	1	O^+	O_3

(a) sp^3 hybridization— Water has the angular structure, with lone pairs of electrons taking up two of the tetrahedral positions, the HOH angle being 104·5°. Thus, comparing the hydrides, CH_4, NH_3, and OH_2, we note that the replacement of one bond by one lone pair of electrons reduces the interbond angles in NH_3 to

107·3°, and the introduction of two lone pairs of electrons in H_2O, in place of two bonds, forces the remaining bonds still closer together. The bond orbitals are described as having an increasing amount of p character along the series CH_4, NH_3, H_2O. We saw in Chapter 10 that in ice, individual water molecules are linked together by hydrogen bonds, a maximum of four hydrogen bonds per H_2O molecule being found at low temperatures. The overall number of hydrogen bonds formed by each water molecule decreases as the temperature is raised, but even in the liquid state at 40°C each molecule forms about half of the theoretically possible number of hydrogen bonds. The tetrahedral arrangement of the hydrogen bonds around the oxygen atom in the ice structure is explained if we assume that the strongest bond is formed along the axis of an oxygen sp^3 hybrid orbital containing a lone pair of electrons (see page 142).

The angular structure is also found for other OX_2 molecules; XOX angles of 103·3°, 110·8° and 111·7° have been reported for F_2O[78], Cl_2O, and OMe_2[79], respectively. In the latter two molecules the bond angles have increased to values slightly greater than the tetrahedral angle (109·5°), and this we attribute to repulsions between either the lone pairs on the chlorine atoms or the bonding pairs in the C—H bonds of the methyl groups.

Oxygen forms a second hydride, H_2O_2, hydrogen peroxide, and x-ray examination of the solid gives the structure shown in *Figure 11.11 (a)* with a H—O—O angle of 96·9°, an 'interplane' angle of 93·6°, and an O—O bond length of 1·49 Å. The crystal structure of the dihydrate, H_2O_2 . $2H_2O$, has also been examined (OLOVSSON and TEMPLETON[80]), and found to fit into the same pattern, with OOH angles of 100·4° and an O—O distance of 1·48 Å; the structure contains planar chains of hydrogen-bonded water molecules crosslinked by hydrogen bonds to the hydrogen peroxide molecules. A possible orbital arrangement for H_2O_2 is sketched in *Figure 11.11 (b)*. The oxygen atoms are shown sp^3 hybridized, with two lone pairs; the O—O bond results from the end-on overlap of an sp^3 orbital from each oxygen atom, and the O—H bonds are formed by the overlap of hydrogen $1s$ orbitals with the remaining sp^3 orbitals on the oxygen atoms. Repulsion between the two lone pairs on each oxygen atom forces the O—H bonds together.

The analogous O_2F_2 molecule (JACKSON[81]) has a similar structure to H_2O_2, the FOO angles being 109·5° and the O—F bond length 1·58 Å. The O—O distance, 1·22 Å, is much less than it is in H_2O_2, and it is apparent that there is considerable multiple bond character. Jackson has proposed that in addition to the O—O σ bond, there is a pair of three-centre orbitals, each embracing one fluorine and two oxygen atoms. The relatively low bond order for the O—F bond then accounts for this bond being much longer than it is (1·41 Å) in OF_2.

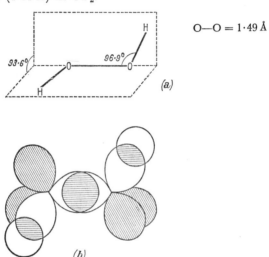

O—O = 1·49 Å

(a)

(b)

Figure 11.11. Structure of H_2O_2. (*a*) Structural parameters; (*b*) Orbital diagram showing the oxygen atoms using sp^3 hybrid orbitals

(b) sp² hybridization—When the oxygen atom has a lone pair and forms one π bond, an angular structure is formed with bond angles approaching the trigonal-planar value (120°). Thus recent microwave studies have shown that ozone, O_3, has the configuration

$$
\begin{array}{c}
\text{O} \\
\diagup \quad \diagdown \\
\text{O} \quad 116 \cdot 8° \quad \text{O}
\end{array}
\quad 1 \cdot 28\ \text{Å}
$$

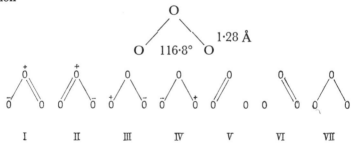

| I | II | III | IV | V | VI | VII |

The valence-bond description of the ozone molecule as a hybrid of the structures I to VII is a clumsy one, and it is better to con-

190

sider that the central oxygen is trigonal planar (sp^2) with one lone pair of electrons and two σ bonds to adjoining oxygens. The simple m.o. diagram (*Figure 9.14*, page 122) proposed for the nitro group may then be used to explain the π bonding. As in the nitro group, there will be two electrons in a bonding three-centre delocalized π orbital and two electrons in a non-bonding π orbital concentrated on the two terminal oxygen atoms.

Sulphur, Selenium and Tellurium

As we go from oxygen to tellurium, we find increasing metallic character, in agreement with the decreasing electronegativity of the elements. Sulphur, selenium and tellurium, moreover, differ from oxygen in their ability to make use of the higher d orbitals, forming up to six covalent bonds (*e.g.* SF_6). Before we summarize possible structures for these heavier elements, we must comment briefly on the higher covalencies, since although the covalent nature of SF_6 is fully established, some workers still doubt whether the sulphur atom can make use of the $3d$ orbitals for bonding; they prefer a valence-bond description for SF_6, in which four bonding electron pairs are distributed between six bonding positions. Two such structures are

It seems preferable to assume that the sulphur atom promotes two of its electrons into the $3d$ levels, and uses sp^3d^2 hybrid orbitals for bonding to the fluorine atoms. MOFFITT[82] has examined the sulphur atom spectrum, and points out that when the atom increases its covalency from 2, to 3, 4 ... *etc.*, the tendency for the electrons to be promoted into $3d$ orbitals is greater than the tendency for them to be lost entirely by ionization. By means of a molecular-orbital treatment, he has shown also that the use of these $3d$ orbitals will give rise to π-bonding in the S—O bonds of the sulphur oxides and related compounds, so that such bonds should be written S=O rather than S→O. This point has been considered in some detail recently by several workers (CRAIG and ZAULI[83], CRUIKSHANK *et al.*[84]) who agree with the proposed incorporation of $3d$ orbitals into the bonding. Readers who are especially interested in bonds formed by oxygen with the second-row elements (Si, P, S) are referred to CRUIKSHANK's paper[85].

Table 11.15 summarizes the structures of compounds formed by these elements.

(a) sp² hybridization—The first type of structure listed in *Table 11.15* is shown by sulphur trioxide, SO_3, which forms a planar molecule in the gas phase, with the three S—O bonds pointing towards the corners of an equilateral triangle. Valence-bond theory (excluding the use of $3d$ orbitals) discusses the molecule in terms of hypothetical structures such as

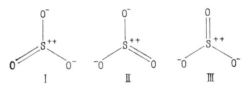

Table 11.15. Structures of S, Se, *and* Te *Compounds*

Hybridiza-tion	Shape	Number of			Oxida-tion state	Examples
		σ bonds	lone pairs	π bonds		
sp^2	Trigonal-planar	3	0	3	VI	SO_3
	Angular	2	1	2	IV	SO_2
sp^3	Tetrahedral	4	0	2	VI	SO_2Cl_2, $[SeO_4]^{2-}$
	Trigonal-pyramidal	3	1	1	IV	$SOCl_2$, $SeOCl_2$
	Angular	2	2	0	II	SCl_2, $Se(CH_3)_2$
sp^3d	Distorted tetrahedral	4	1	0	IV	SF_4, SeF_4, $TeCl_4$
	Trigonal bipyramid	5	0	1	VI	SOF_4
sp^3d^2	Octahedral	6	0	0	VI	SF_6, SeF_6, TeF_6
	Octahedral	6	1	0	IV	$[SeBr_6]^{2-}$, $[TeBr_6]^{2-}$
	Square pyramid	5	1	0	IV	$[TeCl_5]^-$

and this description assigns only partial π-character (one-third) to the S—O bonds. The molecular-orbital treatment of Moffitt, however, shows that the sulphur atom uses its $3d$ orbitals, to give double S=O bonds; the equivalent Lewis structure is

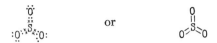

The π bonding results from the lateral overlap of the oxygen p orbitals with sulphur pd hybrid orbitals (rather than pure p or pure d).

We may, of course, describe the π bonding in terms of three delocalized π orbitals compounded from the three oxygen p_z orbitals (xy taken as plane of σ bonds) and the sulphur $3p_z$, $3d_{xz}$ and $3d_{yz}$ orbitals. One such energy scheme is illustrated in *Figure 11.12*. Since the sulphur and oxygen a.o.'s will not correspond

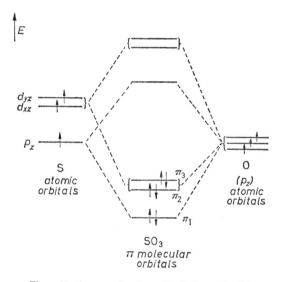

Figure 11.12. π molecular orbital scheme for SO_3

to exactly the same energy, they will be mixed unequally, so the electrons in the π bonding m.o.'s will be centred more over the oxygen than the sulphur atoms.

The length of the S—O bonds in SO_3 (1·43 Å) is exactly the same as that found in the sulphur dioxide molecule, in which the molecule again has a trigonal-planar configuration with a lone pair of electrons occupying one of the positions. The OSO angle in sulphur dioxide (119·5°) is close to the ideal value of 120°, indicating little distortion.

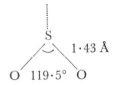

The valence-bond description of the molecule (using no $3d$ orbitals) involves the two forms

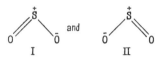

and again implies only partial π bonding (one-half) in the S—O bonds. Since this description gives differing amounts of π character to the S—O bonds in SO_2 and SO_3, it is difficult to see why the bond lengths should be the same, especially as the bond angles (and hence hybridization) are the same. It is evidently more satisfactory to consider the S—O bonds in both molecules to be essentially double, and we can assign the value 1·43 Å to a sulphur-oxygen double bond in a trigonal-planar configuration.

(b) sp³ hybridization—A basically 'tetrahedral' structure is found for compounds in which the Group VI element has four σ electron pairs, irrespective of whether they are bonding pairs or lone pairs. Thus for sulphur we get the tetrahedral arrangement (four σ bonding pairs) for sexivalent compounds of the SO_2X_2 and $[R_3SO]^+$ types, a pyramidal structure (three bonding, one lone pair) for the quadrivalent compounds R_2SO and $[R_3S]^+$, and an angular structure (two bonding, two lone pairs) for the divalent X_2S compounds. The π bonding present in the S—O bonds can be considered (for steric purposes) as merely modifying the tetrahedral angles.

The sulphuryl halides are examples of the sexivalent tetrahedral molecules, and in sulphuryl fluoride, for instance, microwave studies show the OSO and FSF angles to be 124° and 96·1°, respectively; these values would be expected as a result of the repulsions between the S=O bonds (1·41 Å). The crystal structure of sulphuric acid, H_2SO_4, shows that the molecule may be written $SO_2(OH)_2$, in that two of the S—O bonds (1·54 Å) are presumably the S—OH bonds since they are much longer than the others (1·43 Å). The OSO bond angles range from 98° to 117°, the largest angle being that between the two shorter (S=O) bonds. In the sulphate ion, $[SO_4]^{2-}$, and the analogous selenate ion,

$[SeO_4]^{2-}$, all four bonds are equivalent and the bond angles are tetrahedral. The thiosulphate ion, $[S_2O_3]^{2-}$, is structurally analogous to $[SO_4]^{2-}$, with the second sulphur atom replacing one of the oxygen atoms.

Although sulphur trioxide is trigonal planar in the gas phase, it polymerizes in the solid state to give either linear chains (α and β forms) or a cyclic trimer with a puckered ring (γ form)

α and β   γ

In both types of structure, SO_2 groups are linked through S—O—S bonds, and the linking S—O bonds (1·61 Å and 1·60Å for chain and cyclic forms, respectively) are much longer than the terminal S—O bonds (1·41 Å and 1·40 Å, respectively). This is consistent with the view that the linking bonds are single and the terminal bonds double. The configurations are not perfectly tetrahedral, of course, because repulsions between the terminal double bonds will be much greater than those between the linking single bonds; this agrees with the values of 122° and 100° found for the 'outer' and 'inner' OSO angles in the cyclic trimer.

A recent study (MIJCHOFF[86]) of tetragonal selenium trioxide has shown it to be a cyclic tetramer with linking and terminal Se—O of 1·77 Å and 1·55 Å, respectively; the OSeO angles range from 128·2 (outer) to 98° (inner). Good illustrations of the trigonal-pyramidal arrangement (with one lone and three bonding pairs) are provided by the thionyl halides, SOX_2 (X=Br, Cl, F), and by the analogous selenium compound $SeOCl_2$

The sulphite ion, $[SO_3]^{2-}$, also has this pyramidal configuration and it is interesting to note that in the pyrosulphite ion $[S_2O_5]^{2-}$, we get an unsymmetrical structure in which both sulphur atoms are 'tetrahedral', one having a 'sulphite' and the other a 'sulphate' structure

The tellurite ion $[TeO_3]^{2-}$, also has a pyramidal structure (ZEMANN and ZEMANN[87]) with OTeO angles of 100°.

The angular configuration with two lone pairs of electrons is known for a range of sulphur, selenium, and tellurium compounds; bond angles for some of the simpler compounds are quoted in *Table 11.16.*

Table 11.16. Bond Angles for MX_2 *Compounds*

Compound	H_2S	H_2Se	Cl_2S	$(CH_3)_2S$	Br_2Te
Bond angle (°)	92·2	90·8	100·3	98·9	98

It is notable that the bond angles are much smaller for the hydrides than for the other compounds. A reasonable explanation for this has already been given in the discussion of the hydrides of Group VB elements, namely that the size of the atom increases from oxygen to sulphur, which allows the closer approach of the hydrogen atoms before repulsion between the bonding pairs gets important. In the other compounds the bond angles remain around 100° because of repulsions resulting from the lone pairs on halogen atoms or bonding pairs in the methyl groups.

The compounds H_2S_2 and S_2Cl_2 have an S—S bond and are essentially tetrahedral with structures analogous to that of H_2O_2. The same angular arrangement is found for the puckered chains of sulphur atoms in the ions $[S_3]^{2-}$, $[S_4]^{2-}$, $[S_6]^{2-}$, and the rings S_6 and S_8. The S_8 ring is the principal component present in the various allotropic modifications of sulphur in the solid state, and it is also found in the liquid and gaseous states. In this puckered ring, the S—S—S angles are 108·0° and the S—S bond length 2·06 Å (CARON and DONOHUE[88]), and we describe it on the basis of each sulphur atom using four, approximately sp^3 hybridized, tetrahedrally-arranged orbitals. The presence of the lone-pair electrons accounts for the bond angles being less than the tetrahedral value. A similar puckered Se_8 ring is known. The S_6 molecule has a hexagonal (chair form) ring (DONOHUE et al.[89]), with the same S—S distances but somewhat smaller angles (102·2°).

Similar chains of sulphur atoms are found in the polythionates $[O_3S—(S)_n—SO_3]^{2-}$, where n can be zero—giving dithionates—or as large as four, in the hexathionates. A considerable amount of structural work has been done on these compounds in the past few years, especially by Foss[90], who has written an excellent account of the configurations of these compounds. We may briefly summarize the position by saying that there is no branching in the sulphur chain, and that all the sulphur atoms have 'tetrahedral' configurations, the terminal sulphur atoms being bonded to four atoms (three oxygen and one sulphur), and the others having two bonds and two lone pairs. It is interesting to note that the S—S distance depends on the position of the bond in the chain, but otherwise remains constant throughout the polythionates. We may illustrate the point by reference to the hexathionate ion:

The $S_1—S_2$ and $S_5—S_6$ bonds are 2·04 Å and the middle ones $(S_2—S_3, S_3—S_4, S_4—S_5)$ are 2·10 Å. The reasons for this difference are not obvious. It cannot be a direct result of the different valence states of the sulphur atoms (S_1 and S_6 are sexivalent, and S_2, S_3, S_4 and S_5 are divalent), because in the S_8 ring all atoms are divalent and yet the S—S distance is the same as that found for the $S^{II}—S^{VI}$ bonds in the polythionates. For similar reasons, explanations based on differing amounts of partial π bonding seem unlikely. A more promising suggestion is that the differences could arise from differences in the hybridization of the sulphur atoms, and interested readers are referred to a paper on this subject by Lindqvist[91].

(c) *sp^3d hybridization*—Another configuration, which is found for compounds of the quadrivalent elements, is trigonal-bipyramidal, in which one bonding position is occupied by a lone pair of electrons

Both microwave (Tolles and Gwinn[92]) and electron diffraction (Kimura and Bauer[93]) experiments show SF_4 to have this arrangement, the axial S—F bonds (1·65 Å) being significantly longer than the equatorial ones (1·55 Å). Since the equatorial and axial positions are not equivalent, the bond length difference could arise through differences in either the nature of the hybrid σ orbitals or possibly in π bonding. SOF_4, in which sulphur is sexivalent, shows[93] a directly analogous structure, with S—O taking the place of the equatorial lone pair, and once again the axial S—F bonds (1·60 Å) are longer than the equatorial ones (1·54 Å).

Quadrivalent selenium and tellurium also yield trigonal-bipyramidal molecules in their diphenyl and dimethyl dihalides (R_2MX_2), with the phenyl or methyl groups occupying equatorial positions together with the lone pair of electrons. $TeCl_4$ has the same structure as SF_4.

(d) sp^3d^2 hybridization—The octahedral configuration is found for the simple fluorides, SF_6, SeF_6 and TeF_6, as well as for the two substituted ones, SF_5Cl and SF_5Br (NEUVAR and JACHE[94]). S_2F_{10} and $S_2O_2F_{10}$ also have octahedral structures in which SF_5 groups are joined by S—S and S–O–O–S linkages, respectively.

None of the Group VI elements form hexachlorides or hexabromides, but chloride or bromide ions can co-ordinate to the tetrahalides. Thus, in the $[TeCl_5]^-$ ion (AYNSLEY and HAZELL[95]), the five Te—Cl bonds form a square pyramid, with the lone pair of electrons taking up the sixth octahedral position. Hexahalogeno anions, $[MX_6]^{2-}$, are well known, but their octahedral structures would not have been predicted had we considered the lone pair of electrons as stereochemically significant; we should have anticipated a distortion from some seven-co-ordinate configurations (one lone and six bonding pairs of electrons).

Tellurium also achieves an octahedral configuration of oxygen atoms in various tellurates. In $K[TeO(OH)_5]$, H_2O, for instance (RAMAN[96]), there are five Te—O bonds of 1·93 Å and one shorter (1·83 Å).

FLUORINE AND THE GROUP VIIB ELEMENTS

The halogens form diatomic molecules by the mutual overlap of the appropriate p_x orbitals of the two atoms, although this σ bonding may well be supplemented by a small amount of π bonding (MULLIKEN[97]). As we go from fluorine to iodine, so the electronegativity of the halogen atom decreases: fluorine, for instance, is the most electronegative element of all, but chlorine, bromine and iodine form links with more electronegative elements (*i.e.* other halogens and oxygen).

The hydrogen halides, HF, HCl, HBr, and HI, are covalent

gaseous molecules (HF is just liquid at room temperature, b.p. 19·5°C); the bond strength decreases steadily as the halogen atom becomes heavier: H—F = 134, H—Cl = 102, H—Br = 86·5, H—I = 70·5 kcal/mole. The ionic character of the bond decreases along this series, and this decrease in ionic character is correlated with a more symmetrical distribution of the bonding electrons between the hydrogen and halogen atoms. The breaking of *e.g.* the H—Cl bond in a polar solvent such as water to produce H_3O^+ and Cl^- ions, can be pictured as the charge cloud of the bond becoming progressively more concentrated over the chlorine atom as the solvent molecule approaches, until finally the hydrogen atom no longer has any appreciable share of the bonding electrons; H^+ then 'solvates'—combining with a solvent molecule to form H_3O^+. This is clearly shown in the crystal structure of the hydrate of hydrogen chloride, HCl . H_2O, which consists of pyramidal H_3O^+ cations and Cl^- anions.

Hydrogen fluoride has a strongly hydrogen-bonded structure in the solid, liquid, and gaseous states. Solid hydrogen fluoride is found to have a zig-zag arrangement, forming continuous chains throughout the crystal.

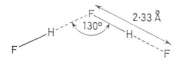

If we consider the fluorine atom to be tetrahedral, with three lone pairs, then the angular nature of the $(HF)_n$ chain can be accounted for (SCHNEIDER[98]), since the hydrogen bonds form along the axis of the lone-pair orbitals. Short zig-zag chains exist in HF vapour, and at temperatures below 60°C polymeric molecules up to $(HF)_5$ can be detected in the vapour by electron diffraction methods.

An angular chain is also found (FORRESTER *et al.*[99]) for the $H_2F_3^-$ ion

Iodine monochloride, ICl, likewise forms zig-zag chains in both its α and β forms (CARPENTER and RICHARDS[100]), the forms differing in whether the Cl branches are *cis* (α) or *trans* (β)

$$\text{Cl}$$
$$|$$
$$--- \text{I---Cl} ----- \text{I} ----- \text{I---Cl} ----- \text{I} ---$$
$$|$$
$$\text{Cl}$$

The bonding has been described in terms of linear three-centre orbitals.

The neutral fluorine atom can form only one bond, but the other halogens can make use of d orbitals and increase their covalency (in the case of iodine to a maximum of seven). With so many possible covalencies, the number of configurations is correspondingly large; these are summarized in *Table 11.17*.

Table 11.17. Configurations of Halogen Compounds

Hybridiza-tion	Shape	Number of			Oxida-tion State	Examples
		σ bonds	lone pairs	π bonds		
sp^3	Tetrahedral	4	0	3	VII	$[ClO_4]^-$, $[IO_4]^-$
	Trigonal-pyramidal	3	1	2	V	$[ClO_3]^-$, $[BrO_3]^-$, $[IO_3]^-$
	Angular	2	2	1	III	$[ClO_2]^-$
	Angular	2	2	0	III	$[ICl_2]^-$
sp^3d	Distorted tetrahedral	4	1	1	V	$[IO_2F_2]^-$
	'T'	3	2	0	III	ClF_3, BrF_3
	Linear	2	3	0	I	$[ClF_2]^-$, $[ICl_2]^-$
sp^3d^2	Octahedral	6	0	1	VII	IOF_5, $[IO_6]^{5-}$
	Square-pyramidal	5	1	0	V	BrF_5, IF_5
	Planar	4	2	0	III	$[BrF_4]^-$, $[ICl_4]^-$
sp^3d^3	Pentagonal-bipyramidal	7	0	0	VII	IF_7

(a) sp^3 hybridization—The perchlorate $[ClO_4]^-$, and metaperiodate $[IO_4]^-$, ions have symmetrical tetrahedral arrangements of oxygen atoms about the halogen, with Cl—O $= 1.50$ Å and I—O $= 1.79$ Å. URCH[101] has suggested that good π bonding is essential if such anions are to be stable, and pointed out that the non-existence of the perbromate ion is to be expected, since the

bromine $4d_\pi$ orbitals would give very poor overlap with the oxygen p_π orbitals, as they have a radial node at the critical overlap position. For the quinquevalent oxyanions, where a lone pair of electrons replaces one of the halogen–oxygen bonds, the remaining bonds can be described as pyramidal, and this arrangement is found in the chlorate $[ClO_3]^-$ bromate $[BrO_3]^-$, and iodate $[IO_3]^-$ (LARSON and CROMER[102]) ions. There is an angular structure (COOPER and MARSH[103]) in the chlorite ion $[ClO_2]^-$ ($\angle$ ClOC= 110·5, Cl—O = 1·56 Å) with two lone pairs of electrons

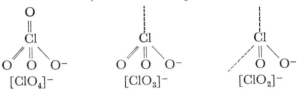

$$[ClO_4]^- \qquad [ClO_3]^- \qquad [ClO_2]^-$$

Chlorine dioxide, ClO_2, also has an angular structure (CURL and HEIDELBERG[104])

$$(\angle \text{ OClO} = 117\cdot4° \text{ and Cl—O} = 1\cdot47 \text{ Å}),$$

but the angle is somewhat larger, and the bond length shorter, than in the chlorite ion.

The π bonding in $[ClO_2]^-$ and ClO_2 can be discussed on the basis of an energy scheme similar to that proposed for the nitro group, with a slight modification allowing for the chlorine atom using a d orbital of suitable symmetry rather than a p_z orbital. WAGNER[105] has discussed the π bond order in a series of chlorine–oxygen compounds and shown that this increases along the series

$$Cl_2O < ClO_2^- < ClO < ClO_3^- < ClO_2 < ClO_4^-;$$

there appears to be a direct correlation between the bond order and the Cl—O force constants.

The angular structure is also found for the $[ICl_2]^+$ ion (VONK and WIEBENGA[106]), which is present in the complexes $[ICl_2][SbCl_6]$ and $[ICl_2][AlCl_4]$, although the angles are now much smaller, as might be expected since there is no multiple bonding in the I—Cl bonds. The ClICl angles differ somewhat in the two complexes, being 92·5° and 96·7°, and this indicates that lattice forces are also influencing the bond angle.

(b) sp^3d hybridization—The halogen atoms (other than fluorine) make use of d orbitals, and while trigonal-bipyramidal structures with five bonding pairs are apparently not found, compounds in which one or more lone pairs of electrons replace bonding pairs are known. Thus $[IO_2F_2]^-$, ClF_3, and $[ICl_2]^-$ have structures in

201

which one, two, and three lone pairs, respectively, are present

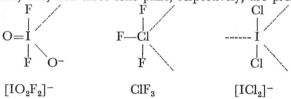

$$[IO_2F_2]^- \qquad ClF_3 \qquad [ICl_2]^-$$

Chlorine trifluoride and bromine trifluoride have similar 'T-shaped' structures; in each case the two apical bonds are longer than the bond in the plane containing the lone-pair electrons, and the FClF and FBrF bond angles are several degrees smaller than 90°. This change in bond angle is again attributed to the repulsion exerted by the lone-pair electrons.

The linear arrangement of the $[ICl_2]^-$ ion (VISSER and VOS[107]) is also found in the analogous ions $[IBrCl]^-$ and $[I_3]^-$. Linear I_3 units can also be weakly linked by iodine molecules to give $[I_7]^-$ and $[I_8]^{2-}$ which have zig-zag chain structures. Thus $[I_8]^{2-}$ has the arrangement

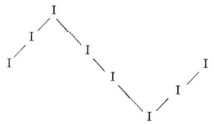

The two I—I distances in the I_3 unit are not always identical, as *Table 11.18* shows.

Table 11.18. I—I *Bond Lengths in the* $[I_3]^-$ *Ion* $[I_{b'}\text{—}I_a\text{—}I_{b'}]$[108]

Compound	I_a—I_b (Å)	I_a—$I_{b'}$ (Å)	I_b—$I_{b'}$ (Å)
NH_4I_3	2·82	3·10	5·92
CsI_3	2·83	3·04	5·87
Cs_2I_8	2·84	3·00	5·84
$(C_2H_5)_4NI_7$	2·91	2·91	5·82
$(C_6H_5)_4AsI_3$	2·90	2·90	5·80

HAVINGA and WIEBENGA[108] and Rundle[109] consider that in the isolated state the I_3^- ion is probably symmetrical, but that deviations arise through anion–cation interactions, the smallest cations having the greatest effect (*cf. Table 11.17*). Other polyhalide ions, $[I(I_2)n]^-$, Br_3^- and Cl_3^- can only be stabilized by large cations. Rundle has developed a molecular-orbital scheme for these polyhalide ions; in the I_3^- case the essential bonding involves a three-centre, σ orbital, just as it did for $(ICl)n$.

(c) sp^3d^2 hybridization—Iodine has an octahedral configuration in the *para*periodate ion $[IO_6]^{5-}$ (FEIKEMA[110]) and probably in IOF_5 (GILLESPIE and QUAIL[111]). When one or two bonding pairs are replaced by lone pairs, we get the square-pyramidal and square-planar configurations, respectively; these arrangements may be illustrated by the species IF_5 and $[ICl_4]^-$ (ELENA *et al.*[112]).

Bromine pentafluoride, BrF_5, has the same configuration as IF_5 (McDOWELL and ASPREY[113]).

The planar arrangement found in $[ICl_4]^-$ is also found in iodine trichloride, I_2Cl_6, which is dimeric with a chlorine-bridged structure

and is almost certainly present in the $[BrF_4]^-$ ion.

(d) sp^3d^3 hybridization—The final bonding arrangement to be discussed for the halogen compounds is the pentagonal-bipyramidal one found in iodine heptafluoride, IF_7

There is considerable controversy over the assignment of this structure to IF_7, which was based on the interpretation of infra-red spectra. Electron diffraction data (LAVILLA and BAUER[114]) fit this structure, provided the five fluorine atoms are not completely planar but staggered, two being somewhat above and two somewhat below the equatorial plane. The available x-ray data have been refined several times by different workers with differing conclusions; the position has been reviewed very recently by DONO-HUE[115] who concludes that the data will not support any model other than the pentagonal bipyramidal one. It is to be hoped that further x-ray studies will be forthcoming to settle this matter.

COMPOUNDS OF THE NOBLE GASES

Despite various early attempts, no chemical compounds of the noble gases were prepared until 1962 and indeed, most chemists dismissed the possibility of their being made. The preparation of a compound formulated as $[O_2]^+[PtF_6]^-$ led BARTLETT and LOH-MANN[116] to consider the feasibility of making the xenon compound, $Xe[PtF_6]$, since O_2 and Xe have almost the same ionization energies. This compound was in fact prepared (BARTLETT[117]), and before long the simple tetrafluoride XeF_4 was also made (CLAASSEN et al.[118]). Since then a great deal of work has been done on various compounds of xenon, and to a lesser extent of krypton, and this has been discussed in a number of textbooks and review articles (e.g. HOLLOWAY[119]). *Table 11.19* summarizes the known structural

Table 11.19. Noble Gas Compounds

Compound	Oxidation state	Shape	I or Te analogue	Reference
XeF_2	II	linear	$[ICl_2]^-$	
XeF_4	IV	planar	$[ICl_4]^{2-}$	
XeF_6	VI	?	—	cf. 110 for original references
$XeOF_4$	VI	square-pyramidal	IF_5	
XeO_3	VI	pyramidal	$[IO_3]^-$	
XeO_4	VIII	tetrahedral	$[IO_4]^-$	120, 121
$[XeO_6]^{4-}$	VIII	octahedral	$[TeO_6]^{6-}$	122, 123
KrF_2	II	linear	—	124

data for the fluoro- and oxo-compounds. The shapes are those predicted on the basis of simple repulsion between pairs of σ electrons (both bonding and lone) except for the hexafluoride. For this compound we should expect a structure based on seven co-ordination (one lone and six bonding pairs), *i.e.* a distorted octahedron, and although the structure has not yet been determined unambiguously, the evidence to date—including a preliminary report on electron diffraction experiments (BARTELL *et al.*[125])—appears to rule out the regular octahedral structure.

The bonding in the noble gas compounds has been the subject of intense discussion, and the matter is by no means settled. Both the valence-bond and molecular-orbital approaches have been used and may be illustrated for xenon (II) fluoride. In the valence-bond procedure, the xenon atom is considered to lose an electron to become Xe^+, and the unpaired $5p_x$ electron then bonds with one or other of the fluorine atoms, *i.e.*

$$F—Xe^+ \quad F^- \qquad F^-Xe^+—F$$
$$I \qquad\qquad\qquad II$$

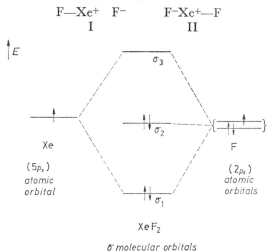

σ molecular orbitals

Figure 11.13. σ molecular orbital scheme for XeF_2

The molecular-orbital method uses the same atomic orbitals ($2p_x$ for F and $5p_x$ for Xe) and compounds them to form three molecular orbitals, one bonding, one non-bonding and one anti-bonding. *Figure 11.13* shows the energy scheme for the σ levels only, and the combinations which give rise to the three molecular orbitals are illustrated in *Figure 11.14*. The combinations (ignoring mixing coefficients) are

$$Fa—Xe—Fb$$
$$\sigma_1 = Fa_\sigma—Xe_\sigma—Fb_\sigma$$
$$\sigma_2 = Fa_\sigma+Fb_\sigma$$
$$\sigma_3 = Fa_\sigma+Xe_\sigma—Fb_\sigma$$

The bonding results from the two electrons present in the σ_1 orbital.

Logical extensions of this molecular orbital approach would use two $5p_z$ orbitals to obtain the square planar bonding in XeF_4,

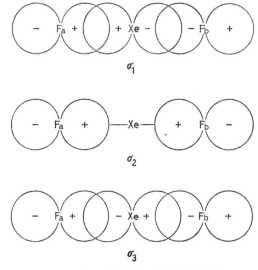

Figure 11.14. Molecular orbitals for XeF_2

and three for XeF_6. The latter would, of course, be octahedral, since the $5s$ orbital contains the lone pair of electrons. Interested readers are referred to recent papers by COULSON[12][6] and LINNETT[12][7] which detail the molecular-orbital approach and shows how it explains the ultra-violet spectra of the fluorides.

REFERENCES

[1] *Tables of Interatomic Distances and Configuration in Molecules and Ions.* The Chemical Society. London 1958, 1965

[2] SCHOMAKER, V. and STEVENSON, D. P. *J. Am. chem. Soc.* 63 (1941) 37

[3] GILLESPIE, R. J. and NYHOLM, R. S. *Q. Rev. chem. Soc.* 11 (1957) 339

[4] PETERS, K. *Z. anorg. Chem.* 131 (1923) 140

[5] BROWN, T. L., DICKERHORF, D. W. and BAFUS, D. A. *J. Am. chem. Soc.* 84 (1962) 1371

[6] WEINER, M., VOGEL, G. and WEST, R. *Inorg. Chem* 1 (1962) 654

[7] WEISS, E. and LUCKEN, E. A. C. *Jnl. organomet. Chem.* 2 (1964) 197

[8] DIETRICH, H. *Acta crystallogr.* 16 (1963) 681

[9] RAHLFS, O. and FISCHER, W. *Z. anorg. Chem.* 211 (1933) 349

[10] RUNDLE, R. E. and LEWIS, P. H. *J. chem. Phys.* 20 (1952) 132

[11] AMIRTHALINGHAM, V., PADMANABHAN, V. M. and SHANKER, J. *Acta crystallogr.* 13 (1960) 201

[12] PETERS, C. H. and MILBERG, M. E. *Acta crystallogr.* 17 (1964) 229

[13] MANN, D. E. and FANO, L. *J. chem. Phys.* 26 (1957) 1665

[14] HEDBERG, K. and RYAN, R. *J. chem. Phys.* 38 (1964) 2214

[15] NORDMAN, C. E. *Acta crystallogr.* 13 (1960) 535

[16] BUNDY, F. P. and WENTORF, R. H. *J. chem. Phys.* 38 (1963) 1144

[17] GROENEWEGE, M. P., SMIDT, J. and DE VRIES, H. *J. Am. chem. Soc.* 82 (1960) 4425

[18] WALLWORK, S. C. and WORRALL, I. J. *J. chem. Soc.* (1965) 1816

[19] FORRESTER, J. D., ZALKIN, A. and TEMPLETON, D. H. *Inorg. Chem.* 3 (1964) 63

[20] GREENWOOD, N. N., ROSS, E. J. F. and STORR, A. *J. chem. Soc.* (1965) 1400

[21] FRASER, G. W., GREENWOOD, N. N. and STRAUGHAN, B. P. *J. chem. Soc.* (1963) 3742

[22] HEITSCH, C. W., NORDMAN, C. E. and PARRY, R. W. *Inorg. Chem.* 2 (1963) 508

[23] PALENIK, G. J. *Acta crystallogr.* 17 (1964) 1573

[24] WOODWARD, L. A., GREENWOOD, N. N., HALL, J. R. and WORRALL, I. J. *J. chem. Soc.* (1958) 1505

[25] GARTON, G. and POWELL, H. M. *J. inorg. nucl. Chem.* 4 (1957) 84

[26] SINNOTT, K. M. *J. chem. Phys.* 34 (1961) 851

[27] LAUNE, V. W., PENCE, D. T. and JACKSON, R. H. *J. chem. Phys.* 37 (1962) 2995

[28] MILLER, R. F. and CURL, R. F. *J. chem. Phys.* 34 (1961) 1847

[29] BRITTON, D. and DUNITZ, J. D. *Acta crystallogr.* 18 (1965) 424

[30] LIVINGSTON, R. L. and RAO, C. N. R. *J. Am. chem. Soc.* 81 (1959) 285

[31] TYLER, J. K. *J. chem. Phys.* 40 (1964) 1170

[32] PARKES, A. S. and HUGHES, R. E. *Acta crystallogr.* 16 (1963) 734

[33] PENFOLD, B. R. and LIPSCOMB, W. N. *Acta crystallogr.* 14 (1961) 589

[34] RUNDLE, R. E. and OLSON, D. H. *Inorg. Chem.* 3 (1964) 596

[35] CARON, A., PALENIK, G. J., GOLDISH, E. and DONOHUE, J. *Acta crystallogr.* 17 (1964) 85

[36] MELLISH, C. E. and LINNETT, J. W. *Trans. Faraday Soc.* 50 (1954) 657

[37] COULSON, C. A. *Valence*, 2nd Ed., p. 221, Clarendon Press, Oxford, 1961

[38] BURNETTE, L. and COULSON, C. A. *Trans. Faraday Soc.* 53 (1957) 403

[39] LIDE, D. R. *J. chem. Phys.* 38 (1963) 456

[40] LIDE, D. R. and MANN, D. E. *J. chem. Phys.* 31 (1959) 1129

[41] BARTELL, L. S. and HIGGINBOTHAM, H. K. *Inorg. Chem.* 4 (1965) 1346

[42] GERMAIN, G., PIRET, P. and MEERSSCHE, M. VAN. *Acta crystallogr.* 16 (1963) 109

[43] TRUTER, M. R., CRUIKSHANK, D. W. J. and JEFFREY, G. A. *Acta crystallogr.* 13 (1960) 855

[44] STERN, S. A., MULLHAUPT, J. T. and KAY, W. B. *Chem. Rev.* 60 (1960) 185

[45] COX, A. P. and RIVEROS, J. M. *J. chem. Phys.* 42 (1965) 3106

[46] KAY, M. I. and FRASER, B. C. *Acta crystallogr.* 14 (1961) 56

[47] KUCZKOWSKI, R. L. and WILSON, E. B. *J. chem. Phys.* 39 (1963) 1030

[48] SMITH, D. W. and HEDBERG, K. *J. chem. Phys.* 25 (1956) 1282

[49] COULSON, C. A. and DUCHESNE, J. *Bull. Acad. r. Belg. Cl. Sci.* 43 (1957) 522

[50] GREEN, M. and LINNETT, J. W. *Trans. Faraday Soc.* 57 (1961) 10

[51] BENT, H. A. *Inorg. Chem.* 2 (1963) 747

[52] MASON, J. *J. chem. Soc.* (1959) 1288

[53] WINNEWISSER, M. and COOK, R. L. *J. chem. Phys.* 41 (1964) 999

[54] LIVINGSTON, R. L. and RAO, C. N. R. *J. phys. Chem.* 64 (1960) 756

[55] GREEN, M. and LINNETT, J. W. *J. chem. Soc.* (1960) 4959

[56] HART, R. R., ROBIN, M. B. and KNEBLER, N. A. *J. chem. Phys.* 42 (1965) 3631

[57] ARNOLD, J. R. *J. chem. Phys.* 14 (1946) 351

[58] VAN WAZER, J. R. *J. Am. chem. Soc.* 78 (1956) 5709; *Phosphorus and Its Compounds*, Vol. 1, Interscience, New York, 1958.

[59] EMERSON, K. and BRITTON, D. *Acta crystallogr.* (a) 17 (1964 1134; (b) 16 (1963) 113

[60] GILLESPIE, R. J. *J. Am. chem. Soc.* 82 (1960) 5978

[61] DAVIES, W. O. and STANLEY, E. *Acta crystallogr.* 15 (1962) 1092.

[62] CRUIKSHANK, D. W. J. *Acta crystallogr.* 17 (1964) 677

[63] AKISTIIN, P. A., RAMBIDI, N. G. and ZASORIN, E. Z. *Kristallografiya* 4 (1959) 360

[64] HAZEKAMP, R., MIGSCHELEL, T. and VOS, A. *Acta crystallogr.* 15 (1962) 539

[65] DOUGILL, M. W. *J. chem. Soc.* (a) (1963) 3211; (b) (1961) 5471

[66] McGEACHIN, H. McD. and TROMANS, F. R. *J. chem. Soc.* (1961) 4777

[67] BULLEN, G. J. *Proc. chem. Soc.* (1960) 425

[68] CRAIG, D. P. *J. chem. Soc.* (1959) 997

[69] DEWAR, M. J. S.   LUCKEN, E. A. C. and WHITEHEAD, M. A. *J. chem. Soc.* (1960) 2423

[70] CRAIG, D. P. and MITCHELL, K. A. R. *J. chem. Soc.* (1965) 4682

[71] HOLMES, R. R., CARTER, R. P. and PETERSON, G. E. *Inorg. Chem.* 3 (1964) 1748

[72] GRIFFITHS, J. E., CARTER, R. P. and HOLMES, R. R. *J. chem. Phys.* 41 (1964) 863

[73] GRIFFITHS, J. E. *J. chem. Phys.* 41 (1964) 3510

[74] MUETTERTIES, E. L., MAHLER, W., PACKER, K. J. and SCHMUTZLER, R. *Inorg. Chem.* 3 (1964) 1298

[75] WHEATLEY, P. J. *J. chem. Soc.* (1964) 2206

[76] GRDENI , D. and ŠĆAVNIČAR, S. *Proc. chem. Soc.* (1960) 147

[77] NORDMAN, C. E. *Acta crystallogr.* 15 (1962) 18

[78] PIERCE, L. and JACKSON, R. *J. chem. Phys.* 35 (1961) 2240

[79] BLUKIS, U., KASAL, P. H. and MYERS, R. J. *J. chem. Phys.* 38 (1963) 2753

[80] OLOVSSON, I. and TEMPLETON, D. H. *Acta chem. scand.* 14 (1960) 1325

[81] JACKSON, R. H. *J. chem. Soc.* (1962) 4585

[82] MOFFITT, W. *Proc. R. Soc.* A200 (1950) 409

[83] CRAIG, D. P. and ZAULI, C. *J. chem. Phys.* 37 (1962) 601, 609

[84] CRUIKSHANK, D. W. J. WEBSTER, B. C. and MAYERS, D. F. *J. chem. Phys.* 40 (1964) 3733

[85] CRUIKSHANK, D. W. J. *J. chem. Soc.* (1961) 5486

[86] MIJCHOFF, F. C. *Acta crystallogr.* 18 (1965) 795

[87] ZEMANN, A. and J. *Acta crystallogr.* 15 (1962) 698

[88] CARON, A. and DONOHUE, J. *Acta crystallogr.* 18 (1965) 563

[89] DONOHUE, J., CARON, A. and GOLDISH, E. *J. Am. chem. Soc.* 83 (1961) 3748

[90] FOSS, O. *Adv. inorg. chem. Radiochem.* 2 (1960) 237

[91] LINDQVIST, I. *J. inorg. nucl. Chem.* 6 (1958) 159

[92] TOLLES, W. M. and GWINN, W. D. *J. chem. Phys.* 36 (1962) 119

[93] KIMURA, K. and BAUER, S. H. *J. chem. Phys.* 39 (1963)3172

[94] NEUVAR, E. W. and JACHE, A. W. *J. chem. Phys.* 39 (1963) 596

[95] AYNSLEY, E. E. and HAZELL, A. C. *Chemy. Ind.* (1963) 611

[96] RAMAN, S. *Inorg. Chem.* 3 (1964) 634

[97] MULLIKEN, R. S. *J. Am. chem. Soc.* 77 (1955) 884

[98] SCHNEIDER, W. G. *J. chem. Phys.* 23 (1955) 26

[99] FORRESTER, J. D., SENKO, M. E., ZALKIN, A. and TEMPLETON, D. H. *Acta crystallogr.* 16 (1963) 58

[100] CARPENTER, G. B. and RICHARDS, S. M. *Acta crystallogr.* 15 (1962) 360

[101] URCH, D. S. *J. inorg. nucl. Chem.* 25 (1963) 2304

[102] LARSON, A. C. and CROMER, D. T. *Acta crystallogr.* 14 (1961) 128

[103] COOPER, J. and MARSH, R. E. *Acta crystallogr.* 14 (1961) 202

[104] CURL, R. F. and HEIDELBERG, R. F. *J. chem. Phys.* 37 (1962) 927

[105] WAGNER, E. L. *J. chem. Phys.* 37 (1962) 751

[106] VONK, C. G. and WIEBENGA, E. H. *Acta crystallogr.* 12 (1959) 859

[107] VISSER, G. J. and VOS, A. *Acta crystallogr.* 17 (1964) 1336

[108] HAVINGA, E. E. and WIEBENGA E. H. *Rec. Trav. chim. Pays-Bas* Belg. 78 (1959) 724

[109] RUNDLE, R. E. *Acta crystallogr.* 14 (1961) 585

[110] FEIKEMA, Y. D. *Acta crystallogr.* 14 (1961) 315

[111] GILLESPIE, R. J. and QUAIL, J. W. *Proc. chem. Soc.* (1963) 278

[112] ELENA, R. J., DE BOER, J. L. and VOS, A. *Acta crystallogr.* 16 (1963) 243

[113] McDOWELL, R. S. and ASPREY, L. B. *J. chem. Phys.* 37 (1962) 165

[114] LAVILLA, R. E. and BAUER, S. H. *J. chem. Phys.* 33 (1960) 182

[115] DONOHUE, J. *Acta crystallogr.* 18 (1965) 1018

[116] BARTLETT, N. and LOHMANN, D. H. *Proc. chem. Soc.* (1962) 115

[117] BARTLETT, N. *Proc. chem. Soc.* (1962) 218

[118] CLAASSEN, H. H., SELIG, H. and MALM, J. G. *J. Am. chem. Soc.* 84 (1962) 3593

[119] HOLLOWAY, J. H. *Prog. inorg. Chem.* 6 (1964) 241,

[120] HUSTON, J. L., STUDIER, M. H. and SLOTH, E. N. *Science, N.Y.* 143 (1964) 1162

[121] CLAASSEN, H. H., CHERNICK, C. L., MALM, J. G. and HUSTON, J. L. *Science, N.Y.* 143 (1964) 1322

[122] IBERS, J. A., HAMILTON, W. C. and MACKENZIE, D. R. *Inorg. Chem.* 3 (1964) 1412

[123] ZALKIN, A., FORRESTER, J. D. and TEMPLETON, D. H. *Inorg. Chem.* 3 (1964) 1417

[124] CLAASSEN, H. H., GOODMAN, G. L., MALM, J. G. and SCHREINER, F. *J. chem. Phys.* 42 (1965) 1229

[125] BARTELL, L. S., GAVIN, R. M. and THOMPSON, H. B. *J. chem. Phys.* 43 (1965) 2547

[126] COULSON, C. A. *J. chem. Soc.* (1964) 1442

[127] BILHAM, J. and LINNETT, J. W. *Nature, Lond.* 201 (1964) 1323

12

COMPLEX COMPOUNDS

IT is difficult to give a formal definition of the term 'complex compound', but we can say that in general it applies to all those compounds in which the number of bonds formed by one of the atoms is greater than that expected from the usual valency considerations. Thus tervalent iron forms six bonds in the complex ion $[Fe(CN)_6]^{3-}$, and bivalent copper forms four bonds in the cuprammonium ion, $[Cu(NH_3)_4]^{2+}$. Of course, there are compounds such as $[C(diars)_2]Br_4$ (COLLINGE *et al.*[2]) which would normally be classified as complex even though the number of covalent or co-ordinate bonds is not different from the group co-valency. We shall use familiar names for complex compounds wherever they can be used unambiguously, but it will often be convenient to adopt the convention whereby Roman numerals, given in parentheses after either the name of the element (in complex cations) or after the name of the complex (in anions), denote the valency of the central atom. Thus the systematic names of $K_3[Fe(CN)_6]$ and $[Cu(NH_3)_4] \cdot SO_4$ become, respectively, potassium hexacyanoferrate(III) and tetramminocopper(II) sulphate. The expression 'complex compound' is used rather than 'co-ordination compound', because it covers a much wider range of substances, and implies no particular bonding characteristics. These compounds may contain either complex anions or cations, or both, or they may be neutral molecules such as the ammine of boron trifluoride, $BF_3 \cdot NH_3$, and *tris*acetylacetoneiron(III), $Fe(acac)_3$. We use the abbreviation '*acac*' to represent the co-ordinating group

$$CH_3 \cdot CO \cdot CH : \overset{.}{C}O \cdot CH_3$$

Abbreviations for some other important organic molecules and groups that form complex compounds are listed in *Table 12.1*.

Such groups are normally referred to as 'Ligands'. When they bond through only one atom (*e.g.* nitrogen in NH_3 and carbon in CN^-), the ligands are called 'monodentate'; groups such as acetyl-acetone and ethylene-diamine, $NH_2 \cdot CH_2 \cdot CH_2 \cdot NH_2$, which can occupy two co-ordinating positions of the metal atom are called 'bidentate' ligands, and the compounds they form are referred to as 'chelate' compounds. Most ligand atoms donate σ lone-pair electrons to the metal ion, although there are some molecules that use π electrons (*e.g.* C_6H_6, C_2H_4).

In general, a metal ion will tend to achieve as high a co-ordination number as possible. The ions of transition metals of the first series are rather small, so that the maximum number of ligands that can be co-ordinated is normally six. With the larger ions of transition metals of the second and third series, however, the maximum co-ordination number increases to eight. Thus, the titanium atom usually forms bonds with only six ligands, but the zirconium atom often bonds to as many as eight ligands. If the ligands themselves are bulky, the co-ordination number may well be lowered.

It is important that co-ordination numbers be only assigned to complexes that are fully characterized, since the composition of the compound may give quite a false impression of the co-ordination number. Thus, the co-ordination numbers of the metal ions in halides such as CuF_2, and MnF_3, are not two and three respectively, since the compounds are polymeric in the solid state with distorted octahedral co-ordinations. In general, therefore, co-ordination numbers should not be assigned unless either molecular weight or x-ray crystal studies have been made. We shall discuss this point under the appropriate co-ordination numbers.

BONDING IN COMPLEX COMPOUNDS

If we consider the hydrates or ammoniates of salts, such as $BaCl_2$, we find that the water, or ammonia, molecules are only weakly held and can be removed easily by heating. These bonds between the metal ions and the water (or ammonia) molecules are termed ion–dipole bonds, because they have their origin in the electro-static attraction of the ion for the polar molecule. Not only do the ammonia and water molecules have permanent dipoles, but additional dipoles are induced by the metal ion, and the resultant electrostatic attraction between the metal ion and the polarized molecule gives rise to ion–dipole bonds. Inasmuch as this attraction is between an ion and a dipole rather than between two ions, the ion–dipole bond is weak, and readily broken, in comparison with an ordinary ionic bond.

Much stronger bonds can be formed, however, if the metal ion has vacant orbitals which can accept electrons from the ammonia or

Table 12.1. Complexing Ligands mentioned in this Chapter

Ligand	Formula of complexing group	Abbreviation for complexing group
Acetylacetone	$CH_3COCHCOCH_3$	*acac*
Ethylenediamine	CH_2NH_2 CH_2NH_2	*en*
Pyridine		*py*
α, α′-Bipyridyl		*bipyr*
Terpyridy		*terpyr*
o-Phenylene-bis-dimethylarsine	$AsMe_2$ ... $AsMe_2$	*diars*
Methyl-bis-(3-dimethyl arsinopropyl)-arsine	$AsMe_2$ $(CH_2)_3$ $AsMe$ $(CH_2)_3$ $AsMe_2$	*triars*
tris-o-Diphenylarsinophenyl-arsine	$\left[Ph_2As \quad \right]_3$ As	*QAS*

water molecules, so forming co-ordinate bonds. It should, perhaps, be emphasized once again that these co-ordinate bonds are essentially covalent bonds, since they involve the sharing of two electrons between two atoms, and that the term 'co-ordinate' merely signifies that the two electrons were originally associated with one atom.

This can be illustrated by further reference to the compound $BMe_3 \cdot NH_3$ (*see* page 71). The bond between the boron and nitrogen atoms is a co-ordinate bond, both electrons coming from the lone pair on the nitrogen atom, and it may be written $\overline{Me_3B}$–$\overset{+}{NH_3}$, showing the theoretical distribution of formal charges once the bond is formed. As boron is less electronegative than nitrogen, the molecular orbital which embraces the boron and nitrogen atoms (forming the B—N bond) becomes distorted (*see Figure 12.1*) so that more of the electronic charge is concentrated near the nitrogen atom, thus tending to neutralize the formal charges.

This simple co-ordination compound is readily formed, because the boron atom has a strong tendency to achieve the symmetrical tetrahedral bonding arrangement, and nitrogen is able to form four bonds by acquiring a formal positive charge. Elements from hydrogen to fluorine in the periodic table can only form up to four co-ordinate bonds since there are only s and p orbitals available, but elements of higher atomic number can form five or six such bonds, since the d orbitals may also be used. We saw in Chapter 9 that the central metal ion might use various hybrid orbitals to give strongly directional bonds, and we can now discuss in detail the orbitals used in the various configurations that are available for

Figure 12.1. The formation of a donor link between BMe_3 and NH_3

each co-ordination number. *Table 12.2*, which summarizes the orbitals used in the more usual configurations, is a simplified form of the table published by KIMBALL[3]. (Unlikely, though theoretically possible, configurations have been omitted.)

The directed valency approach we have so far outlined uses the valence-bond technique, and considers that the bonding electrons belong only to the pair of linked atoms. The molecular-orbital method may also be applied to complex molecules, and a third approach—the 'crystal-field' or 'ligand-field' method—has also been developed for such compounds. Before we discuss the structures of complex compounds systematically, we shall illustrate the way in which all three methods are applied in the case of octahedral complexes, starting with the valence-bond approach.

Table 12.2. Directional Properties of Orbitals

Co-ordination number	Orbitals used	Spatial arrangement of bonds
2	sp or dp	Linear
3	sp^2 or sd^2	Trigonal-planar
4	sp^3 or sd^3	Tetrahedral
	dsp^2	Square-planar
5	dsp^3	Trigonal-bipyramidal
	dsp^3	Square-pyramidal
6	d^2sp^3	Octahedral
7	d^3sp^3 or d^5sp	Face-centred octahedral
	d^4sp^2	Face-centred trigonal-prismatical
	$d^2sp^2 + pd$	Pentagonal-bipyramidal
8	d^4sp^3	Dodecahedral
	d^5p^3	Antiprismatical

Valence-Bond Method

For an octahedral environment, where the ligands are distributed along the x, y, and z axes, we assume that the metal ion makes six orbitals available, the $3d_{z^2}$, $3d_{x^2-y^2}$, $4s$, $4p_x$, $4p_y$, and $4p_z$ (for a first row transition element), and that suitable linear combinations of these are mixed to form six equivalent orbitals (d^2sp^3), (see page 116). The other three $3d$ orbitals, d_{xy}, d_{xz}, and d_{yz}, cannot be used for σ bonding because their lobes point between the axes, but they can be involved in π bonding with ligands that have suitable vacant p or d orbitals. *Figure 12.2* illustrates this d_π—d_π bonding in which a doubly-filled d orbital on M is overlapping a vacant ligand d orbital; since there are only three suitable d orbitals on the metal there is a maximum of three double bonds.

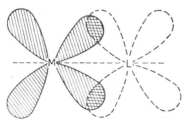

Figure 12.2. Formation of d_π—d_π bonding

We shall discuss the magnetic properties of octahedral complexes later in this chapter, but it may be mentioned at this point that qualitative aspects of the magnetic behaviour of many complexes of transition elements can be accounted for on the basis of a simple valence-bond model. Thus magnetic measurements show that octahedral complexes of tervalent chromium have three unpaired electrons, and we can place one electron in each of the three d orbitals that are not used for σ bonding. However, the method does not completely explain a number of observed magnetic moments. Furthermore, the treatment is only qualitative, and it cannot give even relative energies for the various configurations, so that configurations cannot be predicted. The absorption spectra of complexes cannot be accounted for either, since nothing is known of the energy levels between which the electrons move.

Molecular-Orbital Method (BASOLO and PEARSON[4], GRAY[5])

In this approach the electrons are placed in molecular orbitals which are no longer localized between the metal and any one ligand. By using an LCAO technique we can compound twelve molecular orbitals (six bonding and six antibonding) from the six atomic orbitals of the metal and the six σ orbitals of the ligands; the twelve electrons from the six ligands are placed in the six bonding molecular orbitals. The linear combinations for the six bonding molecular orbitals are as follows:

$$\phi_1 = ad_{z^2} + \sqrt{\frac{1-a^2}{12}} \cdot [2\sigma_z + 2\sigma_{\bar{z}} - \sigma_x - \sigma_{\bar{x}} - \sigma_y - \sigma_{\bar{y}}]$$

$$\phi_2 = ad_{x^2-y^2} + \sqrt{\frac{1-a^2}{4}} \cdot [\sigma_x + \sigma_{\bar{x}} - \sigma_y - \sigma_{\bar{y}}]$$

$$\phi_s = as + \sqrt{\frac{1-a^2}{6}} \cdot [\sigma_x + \sigma_{\bar{x}} + \sigma_y + \sigma_{\bar{y}} + \sigma_z + \sigma_{\bar{z}}]$$

$$\phi_{px} = ap_x + \sqrt{\frac{1-a^2}{2}} \cdot [\sigma_x - \sigma_{\bar{x}}]$$

$$\phi_{py} = ap_y + \sqrt{\frac{1-a^2}{2}} \cdot [\sigma_y - \sigma_{\bar{y}}]$$

$$\phi_{pz} = ap_z + \sqrt{\frac{1-a^2}{2}} \cdot [\sigma_z - \sigma_{\bar{z}}]$$

In these expressions, σ_x and $\sigma_{\bar{x}}$ refer to the σ orbitals of the ligands on the $+ x$ and $- x$ axes respectively, *etc.*, and a is the usual mixing coefficient; thus when $a^2 = 0$ there is no contribution from the d orbitals and the bonding is ionic. In obtaining these linear combinations each metal orbital is taken in turn and combined with all the ligand orbitals with which it can overlap. *Figure 12.3* illustrates the overlap for the ϕ_2 and ϕ_{p_x} molecular orbitals. It can be seen that the $d_{x^2-y^2}$ orbital overlaps equally with all the ligand orbitals in the xy plane, but that no overlap can take place with the z ligands.

Figure 12.4 illustrates the various energy levels concerned in the formation of the σ molecular orbitals.

π-bonding is described by means of π molecular orbitals compounded from the d_{xy}, d_{xz}, and d_{yz} atomic orbitals of the metal and the appropriate p or d orbitals of the ligands. One such bonding π molecular orbital is given by

$$\phi = \beta d_{xz} + \sqrt{\frac{1 - \beta^2}{4}} \cdot [\pi_x + \pi_z - \pi_{\bar{x}} - \pi_{\bar{z}}]$$

where β is the mixing coefficient; when $\beta^2 = 1$ there is no π-bonding and the $3d$ orbitals are described as non-bonding.

Quantitative calculations and predictions about spectra and

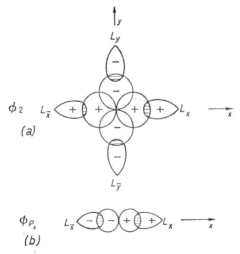

Figure 12.3. (a) Overlap of metal $d_{x^2-y^2}$ orbital with orbitals of ligands (L); (b) overlap of metal p_x orbital with ligand orbitals

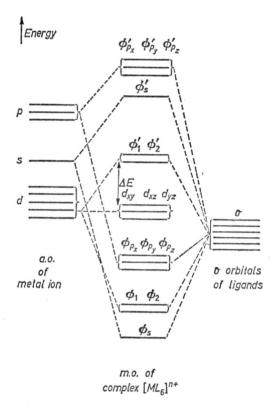

Figure 12.4 Energy levels of molecular orbitals; (degenerate levels are slightly separated and bracketed together for clarity)

magnetic properties can be made by means of the molecular-orbital method, and we shall discuss these matters in more detail later in the chapter.

Ligand-Field Method (BASOLO and PEARSON[4]; ORGEL[6]; COTTON[7])

Early electrostatic models, which attempted to account for the properties of complex compounds on the basis of interactions between point charges and/or dipoles, successfully predicted tetrahedral and octahedral configurations for 4-co-ordinate and 6-co-ordinate compounds, respectively, but could not account satisfactorily for the existence of square-planar complexes. Moreover, it was difficult to see why non-polar ligands such as CO should

218

form stable complexes, or why second and third row elements should form more stable complexes than first row elements (the smaller ions should form the strongest bonds). The theory was subsequently extended to a discussion of the effects of such charges on d levels in the metal ion. This extension, called the 'crystal-field' theory, was originally applied to the behaviour of metal ions in a crystal lattice, but it may be used equally well to describe metal complexes where the metal ion is under the influence of the charge field created by the ligands.

The d orbitals (*see* page 45 for drawings of d orbitals) may be divided into two groups:

(*a*) d_{z^2} and $d_{x^2-y^2}$, which are concentrated along the x, y and z axes. These orbitals are called the e_g orbitals.

(*b*) d_{xy}, d_{xz}, d_{yz}, which are concentrated between the x, y, and z axes. These orbitals are called the t_{2g} orbitals. It is worth noting at this point that the first group of orbitals, the e_g, is that which was used for obtaining bonding orbitals in both the valence-bond and molecular-orbital methods.

If we now consider a typical octahedral complex $[ML_6]^{n+}$, then the σ lone pairs of the six ligands surrounding the ion exert an electrostatic field along the x, y and z axes. Since the e_g orbitals are concentrated along the axes the electrons in these orbitals will be repelled more strongly by the ligands than will the electrons in the t_{2g} orbitals; thus, the t_{2g} orbitals are more stable than the e_g orbitals, and the five degenerate d levels therefore split into an upper doublet and a lower triplet, the energy difference between them being termed Δ (*cf. Figure 12.5*). It is convenient to refer the energies of the d orbitals in ligand-field theory to an energy zero taken as the 'weighted' mean energy of the d orbitals, *i.e.* the energy zero will be nearer to the triplet levels than to the doublet levels, making

$$e_g = +\tfrac{3}{5}\Delta \text{ and } t_{2g} = -\tfrac{2}{5}\Delta$$

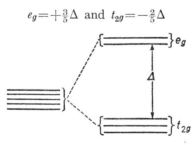

Figure 12.5   Splitting of d levels in an octahedral field

The value of Δ depends upon the charge on the ion, since a large ionic charge causes a large polarization of the ligand electrons and thereby increases the electrostatic field. With the hexahydrates of bivalent ions of the first transition series $[M(H_2O)_6]^{2+}$, for instance, the Δ values are around 10,000 cm^{-1}, as compared with values of about 20,000 cm^{-1} for the analogous tervalent ions. Δ also increases by about 30 per cent on passing from ions of the first transition series to those of the second series. The value of Δ further depends, as might be expected, on the nature of the ligands, since the field produced by the ligands will be related to the ease with which the ligand electrons are distorted by the ion. The ligands may be placed in order of decreasing crystal field effect, an order which is independent of the particular metal ion to which they are attached:

$$CN^- > NO_2^- > NH_3 > H_2O > F^- > Cl^- > Br^- > I^-$$

This order, often referred to as the 'spectrochemical series', may seem a little odd at first, since F^- might perhaps be expected to produce a bigger field than CN^-, but it must be remembered that the electrons are held more tightly to the nucleus of the fluoride ion, and not distorted by the metal ion as much as they are in the 'softer' cyanide ion. In this simple crystal-field approach the forces at work are considered purely electrostatic, and there is no reference to bonding between the metal ion and the ligands, except that which may be implied in the polarization of the lone-pair electrons of the ligands.

Both σ and π bonding, however, may affect the splitting of the d levels, and it is this combination of the ideas of crystal-field and molecular-orbital theory that is referred to as 'ligand-field' theory. Thus the cyanide ion may be considered to produce a large field because it is capable of π-bonding to the metal ion, a process that shortens the metal–ligand distance and increases the field. In the molecular-orbital approach, the value of ΔE depended on the strength of the bonds, since the energy of the antibonding orbitals increases as that of the bonding orbitals decreases. The value of Δ in the ligand-field theory increases as the ligand field increases, whether this arises through electrostatic or through bonding forces. We have referred to the dependence of Δ and ΔE upon the nature of the ligands, and it would perhaps be worth considering briefly how the values of Δ are obtained spectroscopically. We shall do this in the next section, and then discuss the influence of Δ upon the magnetic properties and the stability of octahedral complexes.

ABSORPTION SPECTRA [6, 8, 9]

Compounds of the transition elements with unpaired d electrons are characterized by being coloured, and their absorption spectra accordingly show peaks in the visible and near infra-red region (4000–10000 Å; 25,000–10,000 cm^{-1}). These peaks are of low intensity, their extinction coefficients (ϵ_{max})* being usually between 1 and 50. In addition to these low intensity peaks, the complexes generally show much more intense peaks in the ultra-violet region (2000–4000 Å; 50,000–25,000 cm^{-1}) with ϵ_{max} values of 1000–10,000. If we consider just the first row transition elements for the moment, this division into weak visible peaks and much stronger ultra-violet peaks is surprisingly clear cut.

The weak peaks arise from the transition of electrons from one d level to another, and they are accordingly referred to as d–d transitions. Strictly speaking, these peaks should be of zero intensity, because transitions between two levels of the same l quantum number are 'forbidden' (there being no change of electric dipole). Explanations for the observed low intensities, which involve slight 'mixing' of the d levels with a nearby p level, are discussed in detail in references 6, 8, and 9.

The spectrum of the [TiCl$_6$]$^{3-}$ anion (Russ and Fowles[10]), which shows a broad peak with a maximum at 12,750 cm^{-1} (cf. Figure 12.6), is typical of a d–d transition. Now this anion is an octahedral complex, in which the tervalent titanium has a d^1 configuration, and we can use either the molecular-orbital diagram (Figure 12.4) or ligand-field diagram (Figure 12.5) to explain the spectrum. The latter is the simplest, and using it we say that in the ground state the electron is in a t_{2g} orbital, and may be excited to an e_g orbital. The transition is accordingly referred to as a $t_{2g} \rightarrow e_g$ transition, and 12,750 cm^{-1} represents the splitting Δ. In the molecular-orbital description, ΔE is the equivalent of Δ, so that the peak results from the excitation of the electron from one of the non-bonding d orbitals to one of the lowest-energy antibonding molecular orbitals (either ϕ_1' or ϕ_2'). The spectrum also shows a distinct shoulder, which means that there must be two levels into which the electron may be promoted. The necessary splitting of the e_g doublet may arise because the complex is distorted from the regular octahedron (either because of packing effects or because of the operation of the Jahn–Teller theorem—see later in this chapter).

$$* \; \varepsilon_{max} = \frac{\log I_0/I \times \text{molecular weight}}{\text{concn. (g/l)} \times \text{cell length (cm)}}$$

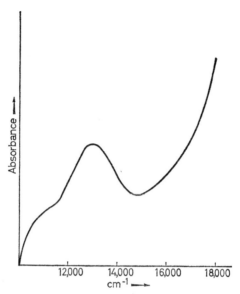

Figure 12.6. The spectrum of the [TiCl₆]³⁻ anion

If we change the environment of the Ti³⁺ ion by surrounding it with six water molecules, then the peak position is observed at 20,300 cm⁻¹, and we can see that a water molecule produces a much greater ligand field than does a chloride ion. In principle, a whole series of tervalent titanium complexes of the type $[\text{TiL}_6]^{3+}$ could be prepared and Δ values determined from the spectra. In practice, it is much easier to make compounds of the type TiX₃, 3L (where $X =$ Cl and Br), and assume that the position of the peak is determined equally by X and L. Experiments of this kind have been carried out recently (FOWLES *et al.*[11, 12]) and ligand-field splittings estimated for a range of oxygen and nitrogen ligands.

Another configuration that is easy to discuss is the d^9 (*i.e.* Cu(II)) in which the ground state is $(t_{2g})^6(e_g)^3$. The peak which is observed results from the promotion of one of the t_{2g} electrons into the e_g orbital that is only half filled. Thus the $[\text{Cu}(\text{H}_2\text{O})_6]^{2+}$ ion shows one broad peak at 12,500 cm⁻¹, and it may be noted that this corresponds to a much smaller splitting than that for $[\text{Ti}(\text{H}_2\text{O})_6]^{3+}$, because the ion carries a smaller charge. In fact all Δ values for hexa-hydrated ions of the first transition series are around 20,000 cm⁻¹ for tripositive ions and 12,000 cm⁻¹ for dipositive ions.

We have discussed the d^1 and d^9 configurations because they are the simplest and because the spectral peaks can be correlated with electron transitions between simple d orbitals. In other d^n ions ($n = 2$–8) the situation is more complicated because interactions between the electrons have to be considered and various 'terms' arise for each configuration. A detailed discussion of the spectra of such ions is beyond the scope of this book and we merely comment that Δ values can be evaluated quite readily from the experimentally observed spectra.

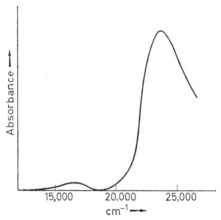

Figure 12.7. The spectrum of $TiCl_3,3C_6H_5N$

In *Figure 12.7* we illustrate the spectrum of $TiCl_3,3C_5H_5N$, and it may be seen that in addition to the d–d peak at 16,600 cm^{-1} there is a much more intense peak at 24,300 cm^{-1} (FOWLES and HOODLESS[13]). This is a 'charge transfer' peak and probably associated with the transition of an electron from a non-bonding titanium d orbital to a pyridine antibonding π^* orbital. We cannot, of course, use the simple diagram of *Figure 12.5* for a discussion of such spectra, but must use a full molecular-orbital diagram. This would resemble *Figure 12.4* but also incorporate the ligand π and π^* orbitals.

We have commented only on spectra of complexes formed by the first transition series because here d–d and charge-transfer transitions may be separated. With the heavier elements Δ (ΔE) gets bigger by at least 30 per cent, as a result of which the d–d peaks occur at lower wavelengths, so that they are often obscured by charge transfer peaks.

All atoms show diamagnetism, (*i.e.* they tend to move from the strongest to the weakest part of an applied magnetic field), but atoms containing electrons with unpaired spins show paramagnetism as well. (They tend to move from a weak to a strong magnetic field.)

We can think of the magnetic moment associated with the 'electron spin' (*see* page 18) as a vector quantity. A pair of electrons with opposed spins will have equal and opposite magnetic moments—thus giving zero resultant moment—but an unpaired electron produces paramagnetism in the parent atom.

The magnitude of the magnetic moment of an atom is, accordingly, related directly to the number of electrons with unpaired spins which are present in the atom, and, if we neglect the complications of azimuthal contribution to the magnetic moment, a simple 'spin-only' formula can be deduced

$$\text{magnetic moment} = \mu = \sqrt{[n(n+2)]} \text{ Bohr magnetons}$$

where n is the number of electrons with unpaired spins and 1 Bohr magneton = 5585·2 erg gauss^{-1} mole^{-1}. This formula is a very good approximation for elements of the first transition series (Sc–Zn).

Any one of the three methods, valence-bond, molecular-orbital, or ligand-field, can be used to discuss the magnetic moments of transition metal complexes, although the latter two are the most satisfactory particularly for quantitative work. If we consider octahedral complexes for metal ions with d^n configurations, for $n = 1$–10, we can place these n electrons in the orbitals available.

As we mentioned earlier, the valence-bond approach used d^2sp^3 hybrid orbitals for the bonding, and puts non-bonding d electrons in the d orbitals not incorporated in the hybridization scheme (*i.e.* d_{xy}, d_{xz}, and d_{yz}). Thus, if we are dealing with the first transition series the hybridization will be $3d^2 4s 4p^3$, and the d electrons will occupy the remaining $3d$ levels. After each level has taken up one electron (*i.e.* d^3), however, the next electron has to pair up—giving a low-spin complex—unless the higher energy $4d$ orbitals are used. It was accordingly suggested (*cf.* BURSTALL and NYHOLM[15]) that two hybridization schemes could be used, $3d^2 4s 4p^3$, or $4s 4p^3 4d^2$, and the complexes using them were referred to as *inner* and *outer* complexes respectively. Theoretical calculations (CRAIG *et al.*[16]) showed that the $4s 4p^3 4d^2$ orbitals would 'stick out' further and, accordingly, overlap at greater distances from the metal atom than

$3d^24s4p^3$ orbitals. The charge cloud of the bonding electrons was, hence, more closely associated with the ligand than with the central atom, giving a greater polarity to the bond. Thus the use of outer orbitals gave covalent bonds that were weaker and more polar than bonds formed from inner orbitals. In general, we notice that outer complexes tend to be formed with the more electronegative ligands such as oxygen and fluorine, and this might be expected because such ligands concentrate the electrons around themselves and will give a better overlap with the more diffuse hybrids involving the outer d orbitals. Inner complexes are commonly formed with ligands of low electronegativity, such as phosphorus and arsenic, especially when d_π—d_π bonding is possible.

Thus tervalent iron forms two series of complexes, (i) with one unpaired electron and (ii) with five unpaired electrons (the configuration of the free Fe^{3+} ion). Typical examples are $[Fe(CN)_6]^{3-}$, $(inner)$ and $Fe(acac)_3$ $(outer)$. The bonding in each general case may be represented as follows:

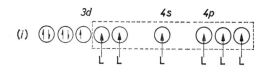

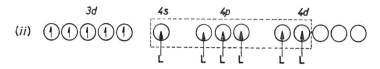

We shall come back to this *inner* and *outer* description later in this chapter when discussing reactivity of complexes.

In the ligand-field approach to magnetochemistry we use the splitting scheme of *Figure 12.5* and apply the principle of putting each electron into the lowest energy orbital that is available, subject to the possible application of Hund's rule (see page 57). Where the ligand field is weak (small Δ), we shall get the maximum number of unpaired electrons, but with a strong ligand field (large Δ) spin pairing will occur. *Table 12.3* lists the magnetic behaviour for the bivalent and tervalent complexes, and *Table 12.4* gives the electron configurations together with the C.F.S.E. for both weak- and strong-field cases.

It may be noted that d^1, d^2, d^3, d^8, and d^9 configurations are

225

precisely the same in both weak and strong ligand fields, but the arrangements of electrons in the t_{2g} and e_g levels are different in the d^4, d^5, d^6, and d^7 configurations. Thus for the d^4 configuration, we can get either the maximum number of unpaired spins (spin free ; high spin)

$$(t_{2g})^3 \ (e_g)^1$$

as, for instance, in the acetylacetone complex of tervalent manganese, $Mn(acac)_3$, or spin-pairing in the t_{2g} levels $((t_{2g})^4)$ to give configuration with only two unpaired electrons such as may be found in the cyanide complex, $[Mn(CN)_6]^{3-}$. The oxygen atoms of the acetylacetone groups produce a relatively weak field and a high-spin complex results, whereas the cyanide ion gives a much stronger field, leading to the formation of a spin-paired octahedral complex.

Table 12.3. Magnetic Behaviour of Octahedral Complexes of bi- and tervalent Transition Elements (for d^n, $n = 4$–7)

| Element | High spin | | Low spin | |
	No. of unpaired electrons	Examples	No. of unpaired electrons	Examples
Cr(II)	4	CrCl$_2$, hydrate	2	K$_4$Cr(CN)$_6$
Mn(III)		Mn($acac$)$_3$		K$_3$Mn(CN)$_6$
Mn(II)	5	[Mn($bipyr$)$_3$]Br$_2$	1	K$_4$Mn(CN)$_6$
Fe(III)		Fe($acac$)$_3$		K$_3$Fe(CN)$_6$
Fe(II)	4	[Fe(en)$_3$]Cl$_2$	0	[Fe($bipyr$)$_3$](ClO$_4$)$_2$
Co(III)		K$_3$CoF$_6$		[Co(NH$_3$)$_6$]Cl$_3$
Co(II)	3	[Co(NH$_3$)$_6$]Cl$_2$	1	[Co($diars$)$_3$](ClO$_4$)$_2$
Ni(III)		—		[Ni($diars$)$_2$Cl$_2$]Cl

The high-spin complexes of ligand-field theory are the same as the outer complexes of the valence-bond approach, and the low-spin complexes correspond to inner complexes. We can summarize the equivalent terms :

(Outer←→spin-free←→high-spin) complex
(Inner←→spin-paired←→low-spin) complex

If we consider the outer high-spin complexes with say d^5 configuration, the two approaches give the same results qualitatively, in that the five d electrons in both cases remain in the d orbitals. The valence-bond method does not differentiate between these d electrons, whereas the ligand-field approach puts them in two

Table 12.4. Electron Arrangements in Octahedral Complexes

Number of d electrons	Weak field (small Δ) t_{2g}	e_g	C.F. S.E. (Δ)	Strong field (large Δ) t_{2g}	e_g	C.F. S.E. (Δ)	Examples
0			0			0	Sc^{3+}
1	↑		0·4	↑		0·4	Ti^{3+}
2	↑ ↑		0·8	↑ ↑		0·8	V^{3+}
3	↑ ↑ ↑		1·2	↑ ↑ ↑		1·2	Cr^{3+}
4	↑ ↑ ↑	↑	0·6	↑↓ ↑ ↑		1·6	Cr^{2+}
5	↑ ↑ ↑	↑ ↑	0	↑↓ ↑↓ ↑		2·0	Fe^{3+}
6	↑↓ ↑ ↑	↑ ↑	0·4	↑↓ ↑↓ ↑↓		2·4	Co^{3+}
7	↑↓ ↑↓ ↑	↑ ↑	0·8	↑↓ ↑↓ ↑↓	↑	1·8	Co^{2+}
8	↑↓ ↑↓ ↑↓	↑ ↑	1·2	↑↓ ↑↓ ↑↓	↑ ↑	1·2	Ni^{2+}
9	↑↓ ↑↓ ↑↓	↑↓ ↑	0·6	↑↓ ↑↓ ↑↓	↑↓ ↑	0·6	Cu^{2+}
10	↑↓ ↑↓ ↑↓	↑↓ ↑↓	0	↑↓ ↑↓ ↑↓	↑↓ ↑↓	0	Zn^{2+}

levels, giving $(t_{2g})^3 (e_g)^2$; the differentiation is essential to explain both the spectra and the distortions observed in some octahedral complexes. It should be noted that this differentiation is also made in the molecular-orbital approach, although the upper doublet is no longer pure d but the antibonding ϕ_1' and ϕ_2' orbitals which are derived from the d_{z^2} and $d_{x^2-y^2}$ orbitals. Thus when the bonding is weak, ΔE (the energy difference between the non-bonding d orbitals and the lowest-energy anti-bonding orbitals) will also be small, and we get the maximum number of unpaired spins. The molecular-orbital treatment can still use the inner d orbitals, however, giving delocalized bonds, whereas the valence-bond approach uses the outer d orbitals to form localized bonds; the outer d orbitals could, of course, be incorporated into the molecular-orbitals.

The simple crystal-field approach merely places the electrons in the t_{2g} and e_g levels, and does not comment on possible covalent bonding between the ion and the ligands. If covalent bonding is also considered (ligand-field), then the outer d orbitals are likely to be concerned in the bonding, together with the s and p orbitals.

When we come to consider the spin-paired complexes, any description will place the first six electrons in the t_{2g} levels. The methods again differ over the placing of subsequent electrons. Thus the valence-bond method places the seventh electron in either a higher s or a higher d orbital (*cf.* Co(II) in *Table 12.4*), whereas

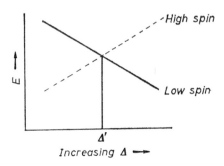

Figure 12.8. Energy relationships for octahedral spin-free and spin-paired complexes

the molecular-orbital and ligand-field methods place this electron in an antibonding molecular orbital and an e_g orbital respectively. The presence of this electron weakens the bonding between the metal and the ligands.

This discussion has shown that whereas the valence-bond method gives a clear-cut division into two classes of compounds, and implies a rather sharp break in the nature of the bonding, the other two methods permit a more gradual transition in bond type. The energy diagram of *Figure 12.8* illustrates this point. For small values of Δ, the spin-free configuration is much more stable than the spin-paired, but as Δ increases so the energy difference gets less, and at a certain critical value (Δ') the two configurations have the same energy. As Δ continues to increase, so the spin-paired state gets increasingly stable relative to the spin-free configuration. Thus although there may be a distinct division on grounds of the number of unpaired electrons, there need not be any such distinction between the types of bonding in spin-free and spin-paired complexes.

THE STABILITY OF COMPLEX COMPOUNDS (*cf.* WILLIAMS[17])

The stability of a complex compound is closely related to its electronic configuration, and depends on the type of bonding which is present. If the bonding is essentially of an ion–dipole character, then we must expect that the strongest bonds will be formed by the smallest ions, since they produce electrostatic fields of the greatest intensity. Considering the hydrated bivalent ions, Mg^{2+}, Ca^{2+}, Sr^{2+}, and Ba^{2+}, and plotting the heats of hydration (a measure of the bond energy) against the reciprocal of the radius, we get the

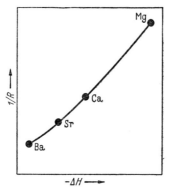

Figure 12.9. A plot of heat of hydration against 1/radius for the bivalent Group IIA ions

smooth curve of *Figure 12.9*, which shows the decreasing strength of the ion–dipole bonds with increasing metal ion radius.

We should expect to get a similar curve for the heats of hydration of the bivalent ions of the elements of the first transition series, since the ionic radii decrease from Ca^{2+} to Zn^{2+} (0·99 Å to 0·72 Å). *Figure 12.10* shows, however, that the heat of hydration does not increase steadily as the ionic radius decreases, the experimental values lying on a curve with two peaks. The hydrated ions other than Mn^{2+} are more stable than would be expected from consideration of relative ionic radii, and this extra stability may be accounted for in terms of crystal field stabilization energies.

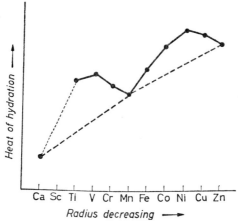

Figure 12.10. Heats of hydration for M^{2+} (Ca—Zn)

229

The ions are octahedrally co-ordinated with six water molecules, $[M(H_2O)_6]^{2+}$, so that the d levels of the metal ion split into the usual t_{2g} triplet and e_g doublet. The hydrates are 'weak-field' complexes (see Table 12.4, page 227), and while the first three electrons enter the t_{2g} levels, conferring stability on the ion, the next two electrons enter the e_g levels and destabilize the ion; electrons 6, 7 and 8 will go into t_{2g} levels and stabilize, whereas electrons 9 and 10 go into e_g levels and destabilize the ion. The deviations in the heats of hydration are the result of the stabilization conferred by the ligand field of the water molecules, and we should expect this stabilization to be greatest for the $d^3(V^{2+})$ and $d^8(Ni^{2+})$ configurations, and to be zero for $d^0(Ca^{2+})$, $d^5(Mn^{2+})$ and $d^{10}(Zn^{2+})$. If the experimental values for the heats of hydration are corrected for the ligand field stabilization (using values of Δ obtained from spectroscopic observations), a smooth curve (shown by the broken line in Figure 12.10) is obtained. Similar effects are noted for the hydrates of tervalent ions, with maximum stabilization for the d^3 configuration (Cr^{3+}) and no stabilization for $d^5(Fe^{3+})$.

The stability of complexes in solution can also be discussed in terms of ligand field effects. Here we are concerned with a reaction in which one or more water molecules of the hexahydrated ion is replaced by another ligand, e.g.:

$$[M(H_2O)_6]^{2+} + X \rightleftharpoons [M(H_2O)_5X]^{2+} + H_2O$$

If the equilibrium constant (often called the 'stability constant' in this context) for this reaction is K, then log K, which gives the free energy change for the reaction, is a measure of the stability of the complex hydrate. Many stability constants have been recorded,[18] but we shall limit our discussion to ethylenediamine (en) complexes (GRIFFITH and ORGEL[19]). Each ethylenediamine molecule (a bidentate ligand) can replace two water molecules, and the stability constants K_1, K_2 and K_3 for the following reactions have been determined:

$$[M(H_2O)_6]^{2+} \quad + en \rightleftharpoons [M(H_2O)_4\,en]^{2+} + 2H_2O \quad \ldots \quad K_1$$
$$[M(H_2O)_4\,en]^{2+} + en \rightleftharpoons [M(H_2O)_2\,en_2]^{2+} + 2H_2O \quad \ldots \quad K_2$$
$$[M(H_2O)_2\,en]^{2+} + en \rightleftharpoons [Men_3]^{2+} \qquad\quad + 2H_2O \quad \ldots \quad K_3$$

The sum (log K_1 + log K_2 + log K_3) gives a measure of the stability of $[M\,(en)_3]^{2+}$ compared with $[M(H_2O)_6]^{2+}$, see Table 12.5. The 'corrected' values shown in Table 12.5, which allow for the stabilization effects, increase smoothly from Mn^{2+} to Zn^{2+}.

Table 12.5.   Stability of $[M \ (en)_3]^{2+}$

	Mn²⁺	Fe²⁺	Co²⁺	Ni²⁺	Cu²⁺	Zn²⁺
log K_1 + *log* K_2 + *log* K_3	5·67	9·52	13·82	18·06	18·60	12·09
log K_1 + *log* K_2 + *log* K_3 (*corrected for C.F.S.E.*)	5·67	6·95	8·24	9·52	10·80	12·09

Similar considerations of C.F.S.E. have been used to explain variations in the energy relationships of a number of types of octahedral complexes of the first transition series. GUZZETTA and HADLEY,[20] for instance, have used direct calorimetry to measure the heats of complexation of cyanide ion with number of dipositive ions, V²⁺, Cr²⁺, Mn²⁺, Fe²⁺, Co²⁺, Zn²⁺, and from measurements of the heats of combustion and sublimation of acetylacetone complexes, M(*acac*)₃, for M = Sc, Ti, V, Cr, Mn, Fe, and Co, WOOD and JONES[21] have calculated metal-ligand bond energies. In each case the variations observed were accounted for by allowing for C.F.S.E. Readers interested in the general application of these ideas to variations in the thermodynamic properties (including lattice energies) of complexes are referred to an excellent article by GEORGE and McCLURE.[22]

Discussions of the type we have outlined here for octahedral complexes can be extended to complexes of other co-ordination numbers; thus BERG and SINANOĞLU[23] have discussed the stability of divalent oxides and halides of the transition metals which exist as discrete linear molecules at high temperatures.

DISTORTION OF OCTAHEDRAL COMPLEXES

Up till now we have assumed that in a complex ML_6 all the bond lengths and bond angles are identical, and this is true for complex ions such as $[TiCl_6]^{2-}$, although we might expect some deviations from 90° angles when the six ligands are not equivalent or when polydentate ligands are involved. We might, for instance, expect the angles in a chelate to depend upon the size of the ring.

It is necessary for us to consider one very important additional factor, however, that of repulsion between the ligand electrons and any d electrons on the central metal ion. We have seen already that the d electrons can be considered under two categories, those which point towards the ligands (e_g) and those which point between the ligands (t_{2g}). If the e_g levels are unevenly filled then we should expect some of the ligands to experience a bigger repulsion than

231

others, and the metal-ligand bond lengths to differ. Inequality in the t_{2g} levels, on the other hand, should have much less effect, and only minor distortions would be expected. *Tables 12.6 and 12.7* list the predicted distortions for high-spin and low-spin complexes respectively.

High-spin Complexes (cf. Table 12.6)

There should be no distortion for the configurations d^0, d^3, d^5, d^8, and d^{10}, since in each case the charge density is spherically symmetrical, and the d^1, d^2, d^6, and d^7 configurations should not give rise to serious distortions because the asymmetry is in the t_{2g} levels. We should, however, expect to find serious distortions in the d^4 and d^9 complexes with configurations $(t_{2g})^3 (e_g)^1$ and $(t_{2g})^6 (e_g)^3$ respectively. If, for the d^4 complexes we consider that the e_g electron is in the d_{z^2} orbital, then we would expect the complex to have four short and two long bonds (z axis)—a 4/2 system; putting the electron in the $d_{x^2-y^2}$ orbital would give two short and four long bonds—a 2/4 system. There is no reason to expect either a 4/2 or a 2/4 system to be preferred, but in practice it is the 4/2 structure that predominates.

The complex fluorides KMF_3, for M = Cr, Mn, Fe, Co, Ni, and Cu, provide (KNOX[24]) an interesting series, since they do not contain discrete $[MF_3]^-$ anions but are polymeric through M—F→M bridging which gives a six-co-ordinate environment to the metal atoms. The Mn, Fe, Co, and Ni compounds contain metal atoms with a regular octahedral arrangement of fluorine atoms, but both the Cr and Cu compounds have tetragonal structures with two M—F distances appreciably longer than the other four. A similar effect is observed for the series of compounds $MCl_2,2H_2O$ (*cf. Table 12.8*). Thus the Mn, Fe, and Co complexes consist of polymeric planar chains of MCl_2 units with the two water molecules taking up the remaining two octahedral positions with 'normal' M—O distances.

In the analogous copper compound the structure consists of planar $CuCl_2(H_2O)_2$ units connected through longer Cu—Cl bonds.

Table 12.6. Predicted Distortions for High-Spin Octahedral Complexes

No. of d electrons	Configuration		Predicted distortion
	t_{2g}	e_g	
0	0	0	None
1	1	0	Very small
2	2	0	Very small
3	3	0	None
4	3	1	Appreciable
5	3	2	None
6	4	2	Very small
7	5	2	Very small
8	6	2	None
9	6	3	Appreciable
10	6	4	None

Table 12.7. Predicted Distortions for Low-Spin Octahedral Complexes

No. of d electrons	Configuration		Predicted distortion
	t_{2g}	e_g	
4	4	0	Very small
5	5	0	Very small
6	6	0	None
7	6	1	Appreciable

Table 12.8. Bond Lengths in the $MCl_2,2H_2O$ Complexes

Metal	M—O (Å)	M—Cl (Å)	Reference
Mn	2·15	2·52; 2·59	(25)
Fe	2·08	2·49; 2·54	(25)
Co	2·04	2·45; 2·48	(26)
Cu	2·01	2·31; 2·98	(27)

The simple dihalides and trihalides are also polymeric with 'octahedrally' co-ordinated metal atoms, and the d^4 and d^9 metal atoms show typical 4/2 structures. Examples of such distorted octahedra include $CrCl_2$ (TRACY *et al.*[28]), MnF_3 (HEPWORTH and JACK[29]) and the cupric halides, CuX_2 (X = F, Cl, and Br) (*cf.* ORGEL[27]). An especially interesting compound is Cr_2F_5 (STEIN-FINK and BURNS[30]) which contains equal amounts of two types of chromium atom, one (Cr(III)) with a regular and one (Cr(II)) with a distorted octahedral environment of fluorine atoms.

These distortions are illustrations of the general Jahn-Teller theorem, which says that if a non-linear molecule has a degenerate state, then there is at least one vibrational co-ordinate along which a distortion can occur so as to remove the degeneracy. Thus in an octahedral arrangement, the four ligands in the xy plane can move towards the metal ion, and simultaneously the two ligands on the z axis move away, so that the $d_{x^2-y^2}$ orbital is destabilized and the

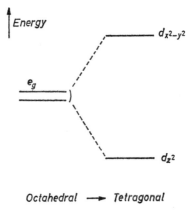

Octahedral ⟶ Tetragonal

Figure 12.11. Splitting of e_g doublet by tetragonal distortion of octahedral complex

d_{z^2} stabilized. This distortion continues until the energy gained (by putting the odd e_g electron in a lower-energy orbital—the d_{z^2}) is just balanced by the energy required to compress and stretch the bonds. *Figure 12.11* shows this splitting of the e_g doublet. We shall return to this aspect of the distortion of octahedral complexes when we consider square planar compounds (*see* page 244), because the latter can be thought of as the limiting case of the two z axis ligands being removed so far away as not to be exerting a significant ligand field on the metal ion.

It is worth commenting that the molecular-orbital approach also explains these distortions, because the d_{z^2} electron is now placed in the ϕ_1' antibonding orbital which is still concentrated along the z axis and will accordingly weaken the bonds to the ligands directed along the $\pm z$ axis. It is possible to give an explanation by means of the valence-bond theory, if we allow two different modes of

234

hybridization for the metal ion—$3d4s4p^2$ for the four short bonds in the xy plane, and $4p_z4d_{z^2}$ for the two longer bonds. However,

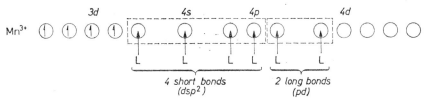

this is merely a convenient explanation to fit the facts, and it is questionable whether useful predictions could be made by this approach.

We might have anticipated a similar distortion in the acetyl-acetone complex $Mn(acac)_3$, but a recent x-ray study (MOROSIN and BRATHOVDE[31]) shows that the distortion is in the bond angles rather than bond lengths.

Low-Spin Complexes (cf. Table 12.7)

There should be no distortion for d^6 complexes, and the effect in d^4 and d^5 complexes should be fairly small. In d^7 complexes, on the other hand, where there is just one electron in the e_g levels, we would anticipate distortions analogous to those found for the high-spin d^4 complexes. Unfortunately, no x-ray investigations have yet been carried out to test this suggestion concerning d^7 complexes.

Up till now we have assumed that d^8 complexes have the $(t_{2g})^6 (e_g)^2$ configuration, with one electron in each of the e_g orbitals. This is generally true, but recently several six-co-ordinate dia-magnetic d^8 complexes of divalent nickel, palladium, and platinum have been reported. The diarsine complexes $M(diars)_2I_2$ are typical examples (STEPHENSON[32]).

On the ligand-field approach we consider these electrons to be in one of the e_g levels, and apply the Jahn-Teller theorem, so that the e_g levels split with the two electrons now pairing up in the lower level. We accordingly predict that there should be an appreciable distortion from the ideal octahedral configuration in these com-pounds.

235

The diarsine complexes are *trans* with the bond distances quoted in *Table 12.9*.

Table 12.9. Bond Lengths in $M(diars)_2I_2$ Complexes

Metal	M—I (Å)	M—As (Å)
Ni	3·21	2·29
Pd	3·40	2·39
Pt	3·50	2·38

The M—I distances are considerably longer than expected for single bonds (the Pd—I distance in a typical trans square-planar complex is 2·65 Å), and it would seem that the pair of electrons is occupying the d_{z^2} orbital and strongly repelling the two iodine atoms. The M—As distances, on the other hand, are about 0·1 Å shorter than expected, and this has been attributed to a measure of d_π—d_π bonding.

CO-ORDINATION NUMBERS OTHER THAN 6

The three methods outlined earlier for octahedral complexes, namely valence-bond, molecular-orbital, and ligand-field, can be used to describe complexes of other co-ordination numbers. The application of the first two methods to simple compounds such as $BeCl_2$ (linear), BCl_3 (trigonal-planar), and CH_4 (tetrahedral) was discussed in Chapter 9 in some detail, and very little modification is needed when the theories are applied to complex compounds with these stereochemistries, apart from the possible incorporation of d orbitals into the bonding descriptions.

With co-ordination numbers 2 and 3 the known complexes are formed by ions with d^{10} configurations, so that the ligand-field approach is of no value. Mainly the simple valence-bond approach is used for co-ordination numbers 5, 7, and 8, where relatively few compounds are known, but for co-ordination number 4 the ligand-field description will be discussed at some length and compared with the valence-bond one.

CO-ORDINATION NUMBER 2

The only well characterized complex compounds of co-ordination number two are those formed either by the univalent ions of silver and gold (with a d^{10} configuration), or by ions with a d^0 configuration such as uranium in $[UO_2]^{2+}$; these compounds all have linear structures. Thus, in the complex cyanide ions $[\overset{-}{N}\overset{-}{C} \to \overset{+}{M} \leftarrow \overset{-}{C}N]^-$, where M = Ag, Au, the two co-ordinated cyanide groups and the metal ion are collinear and the same linear arrangement is probably present in the complex ammines, $[M(NH_3)_2]^+$. The crystalline solid $KCu(CN)_2$, however, is not a 2-co-ordinated complex of Cu^+; it contains, as we shall see in the next section, polymeric spirals of linked copper atoms with a co-ordination number of three. Some other apparently 2-co-ordinate complexes, e.g. $CuI \cdot CH_3NC$ (FISHER et al.[33]), are polymeric, and these contain 4-covalent copper atoms.

One interesting compound studied recently (BROWN and DUNITZ[34]) is diazoaminobenzene copper(I), Cu(PhNNNPh), which has an essentially planar structure with the copper atoms having linear co-ordination. The Cu—Cu distance is quite short (2·45 Å).

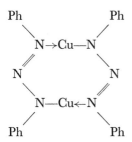

The 2-co-ordinate complexes are diamagnetic (the metal ion having the d^{10} configuration), and they can most simply be described by valence-bond theory, using sp hybrid orbitals. Thus we can write the orbital configuration of the Ag^+ ion in the complex cyanide ion as

where the cyanide ions donate electron-pairs into the vacant sp hybrid orbitals ($5s5p$ for silver, $6s6p$ for gold) forming two σ bonds. The simple molecular-orbital approach outlined in Chapter 9 for beryllium dichloride can, of course, be used with advantage.

The linear configuration is also found in the cyanide and thiocyanate of univalent silver. These compounds have polymeric chain structures,

$$-Ag-C\equiv N\rightarrow Ag-C\equiv N\rightarrow Ag-$$

and it is interesting to note here that in the cyanide, where the linking is through carbon and nitrogen, the chains are straight, whereas the thiocyanate has a zig-zag structure, because of the tetrahedral bond configuration around the sulphur atoms (*see* Chapter 11, page 194).

A linear arrangement is also observed in simple compounds of mercury such as the halides (HgX_2, X = Cl, Br and I) and alkyl halides (*e.g.* CH_3HgX, X = Cl and Br), and, to a lesser extent, in the halides of zinc and cadmium (ZnI_2, $CdBr_2$ and $CdCl_2$). We have to account for this greater prevalence of co-ordination number two amongst compounds of the bigger and heavier elements, and in a very simple way we can consider that the d^{10} charge clouds of the heavier atoms are deformed relatively easily on the approach of the ligands. Thus the two ligands would tend to approach from opposite directions, say along the z axis, and the d^{10} charge cloud would be distorted so as to increase the charge density in the xy plane. This increased charge density helps to prevent the close approach of other ligands. ORGEL[35] has described this redistribution of charge in terms of hybridization of the s and d_{z^2} orbitals (*see Figure 12.12*):

$$\psi_1 = \psi_s + \psi d_{z^2}$$
$$\psi_2 = \psi_s - \psi d_{z^2}$$

ψ_2 is occupied by the two electrons originally considered to be in the d_{z^2} orbital. It must be emphasized, however, that this hybridization approach is merely a way of looking at the charge cloud reorientation and does not describe the bonding between the metal and the ligands.

We can, however, make use of the unoccupied orbital (ψ_1) in

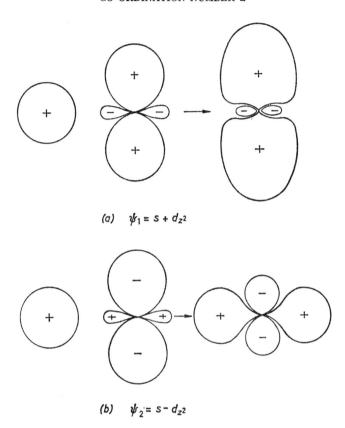

(a) $\psi_1 = s + d_{z^2}$

(b) $\psi_2 = s - d_{z^2}$

Figure 12.12. sd hybridization

a simple molecular scheme for the bonding, by combining it and the p_z orbital with the two ligand orbitals. The resultant energy level diagram, which is shown in *Figure 12.13*, closely resembles that for $BeCl_2$ (*cf. Figure 9.5*) except that the *sd* hybrid replaces the *s* orbital.

We should expect a collinear arrangement of bonds in the complex ions $[UO_2]^{2+}$, $[VO_2]^+$ and $[MoO_2]^{2+}$, in which the metal atom can formally be considered as having the configuration d^0. Thus we can write $[VO_2]^+$ as $V^{5+}(O^{2-})_2$, with the doubly-charged oxide ions each donating two electrons into vacant *sp* hybrid orbitals of V^{5+}. However, a more satisfying description is

$$[O=\overset{+}{V}=O]^+,$$

239

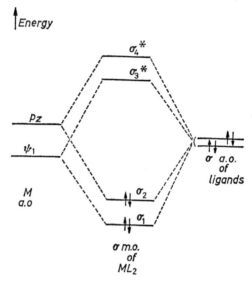

Figure 12.13. Molecular-orbital energy diagram for linear ML_2 complexes

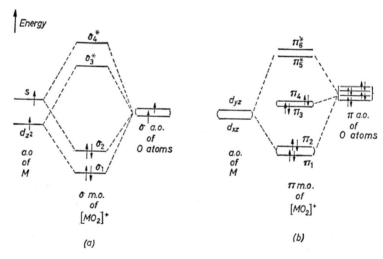

Figure 12.14. Molecular-orbital energy diagrams for $[MO_2]^+$
(a) σ (b) π

with σ bonds arising from sp or pd hybrid orbitals, and π bonds arising from the overlap of singly-filled d orbitals with singly-filled p orbitals on the oxygen atoms.

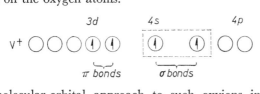

The molecular-orbital approach to such oxyions incorporates the s and d_{z^2} atomic orbitals for σ bonding (along the z axis) and the d_{xz} and d_{yz} orbitals for π bonding. The σ and π orbital schemes are shown separately for clarity in *Figures 12.14(a)* and *12.14(b)*. The two bonding π molecular orbitals will be similar to those proposed for CO_2 (*see* page 170) and be distributed over all three atoms. Non-linear complex compounds of co-ordination number two are unusual; one example is the $[ICl_2]^+$ ion discussed in Chapter 11, page 201. There we saw that ions of this type were best considered as essentially tetrahedral, with lone-pair electrons occupying two bonding positions.

CO-ORDINATION NUMBER 3

The predicted bonding arrangement (using sp^2 or sd^2 hybrid orbitals) is 'trigonal-planar'. This configuration is known in simple molecules such as BCl_3, as we have seen in Chapter 11, but many molecules with structures apparently of this kind are really more complicated; aluminium chloride, for example, is Al_2Cl_6, and not $AlCl_3$. Similarly, the salts $MCuCl_3$ ($M=NH_4$, Li, and K) (Vossos *et al.*,[36] WILLETT *et al.*[37]) do not contain $[CuCl_3]^-$ anions but planar dimers $[Cu_2Cl_6]^{2-}$. The analogous gold compound $CsAuCl_3$ contains a mixture of linear gold (I) ($[AuCl_2]^-$) and square-planar gold (III) ($[AuCl_4]^-$) anions. Co-ordination number three is, in fact, quite rare in complex compounds.

One example appears to be $Me_2Be \leftarrow NMe_3$, which is made by the direct reaction of beryllium dimethyl and trimethylamine. Vapour pressure studies show this molecule to be monomeric in the vapour state, although it seems likely that dimers will be present in the solid. Apart from this beryllium compound, 3-co-ordinate complexes seem to be formed mainly by the d^{10} ions of univalent copper, silver and gold and of divalent zinc. As we mentioned in the last section, $KCu(CN)_2$ contains spiral polymeric chains in which the linked copper atoms take up a co-ordination of three.

The C—Cu—C angle is somewhat greater than the expected 120°. A similar trigonal-planar arrangement is found in $KCu_2(CN)_3$, H_2O where the structure consists of polymeric sheets of $[Cu_2(CN)_3]_\infty^-$ units containing linked $(CuCN)_6$ puckered hexagons. In $Na[Zn(OH)_3]$, the zinc atom has a trigonal planar environment of OH groups (SCHNERING[38]).

A number of 3-covalent tertiary phosphine complexes of univalent copper and silver iodide have been characterized (LANE and PAYNE[39]; CASS et al.[40]), e.g.

$$((Et_2N)_2PhP)_2CuI, \qquad (Me_2N \cdot C_6H_4 \cdot PMe_2)_2AgI,$$
$$(CF_3 \cdot C_6H_4 \cdot PEt_2)_2AgI,$$

and

$$(PhPMe_2)_2AgI.$$

These compounds are monomeric in solution, and they are presumably trigonal-planar, the metal atoms using sp^2 hybrid orbitals.

It is not clear why 3-co-ordination arises when phosphorus ligands are used rather than nitrogen. ARHLAND and CHATT[41] suggested that it could be because phosphorus is capable of d_π—d_π bonding, and that this bonding is much more effective for trigonal-planar than tetrahedral bonding, but COTTON[42] has disputed the latter supposition. The recent preparation of tetrahedral complexes $[ML_4]ClO_4$ (M = Cu, Ag, and Au) with the ligand suggests that the explanation is a steric one, and that only with bulky tertiary phosphines is the co-ordination limited to three.

We can discuss trigonal-planar stereochemistry along the lines proposed for linear complexes, and consider that as the three ligands approach the metal atom they will be as far apart from one another as possible in, say, the xy plane. The consequent deformation of the d^{10} charge cloud gives an increased charge density along the

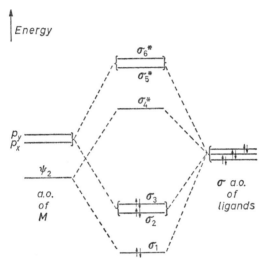

Figure 12.15. Molecular-orbital energy diagrams for trigonal-planar ML_3 complexes

z axis, which may be described by saying that the d_{z^2} electrons now occupy an sd hybrid orbital (that given by ψ_1 on page 238). The bonding can be described on a simple molecular basis by using the other sd hybrid (ψ_2) and the p_x and p_y orbitals of the metal together with an orbital from each ligand. *Figure 12.15* shows the simple molecular-orbital diagram, and it may be seen that this closely resembles that given for BCl_3 (*Figure 9.7*) except that an sd hybrid replaces the s orbital used for boron.

We have already seen in Chapter 11 that pyramidal complexes can be formed by positively-charged oxygen and sulphur atoms (*e.g.* in $[H_3O]^+$ and $[Me_3S]^+$), but these structures are better considered as tetrahedral, with a lone-pair of electrons occupying one of the four positions.

CO-ORDINATION NUMBER 4

The principal stereochemistries observed experimentally are tetrahedral and square-planar, although various distorted versions of these two structures may be found when all the ligands are not the same or if the d levels are filled unevenly. If we limit our discussion for the moment to undistorted complexes, then the valence-bond description of tetrahedral bonding involves sp^3 or sd^3 (d_{xy}, d_{xz}, d_{yz}) hybridization schemes, or a mixture of the two; square-planar hybridization arises from the combination of $d_{x^2-y^2}$, s, p_x and p_y atomic orbitals, the resulting hybrids having axes lying in the xy plane.

The tetrahedral arrangement is found for complexes of divalent beryllium, zinc, cadmium, and mercury, and of tervalent boron, aluminium, and gallium. These elements have no d orbitals comparable in energy with the valence shell s and p orbitals, so that if the co-ordination is restricted to four then a tetrahedral arrangement is to be expected. (With some of these metals the co-ordination number can be increased to 5 or 6 through the incorporation of the high energy d orbitals into the bonding scheme—e.g. $AlH_3,2NMe_3$). If, however, d levels are energetically available, then the 4-co-ordinate complex may be either tetrahedral or square-planar. It is not possible to predict from valence-bond calculations which configuration will have the lowest energy for a given metal ion and ligands, and most quantitative work has been based on either the molecular-orbital or the ligand-field methods.

The molecular-orbital method uses the same orbitals as the valence-bond approach, but compounds them with ligand orbitals to give molecular orbitals. The orbital symmetries are such that the $d_{x^2-y^2}$ and s orbitals of the metal can combine with orbitals on all four ligands, but the p_x and p_y orbitals can only overlap with the ligands orientated along the x and y axes respectively.

We shall describe the ligand-field theory of 4-co-ordination in rather more detail since it has been used extensively to describe these complexes. *Figure 12.16* illustrates a tetrahedral bond configuration, and shows that the x, y and z axes bisect the angles between pairs of ligands. If we now apply the argument already outlined on page 219 for octahedral complexes, we see that the d_{z^2} and $d_{x^2-y^2}$ orbitals of the metal ion are further away from the ligand electrons than are the electrons in the d_{xy}, d_{yz} and d_{zx} orbitals. In this case, therefore, the effect of the ligand field is to split the d levels, giving a stabilized doublet level, e, of lower energy than the undistorted d orbitals, and less stable triplet levels, t_2, of higher

244

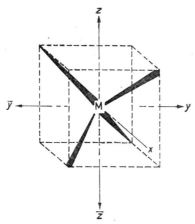

Figure 12.16. Tetrahedral bond configurations

energy (*see Figure 12.17(a)*). The field produced at the metal ion by four tetrahedrally arranged ligands is, however, smaller than that produced by six of the same ligands at the same distance in an octahedral configuration; thus, if Δ is the splitting for octahedral co-ordination, the value for tetrahedral co-ordination is considered to be about $\Delta\frac{4}{9}$. We saw on page 219 that the t_{2g} and e_g levels in an octahedral field have energies of $-\frac{2}{5}\Delta$ and $+\frac{3}{5}\Delta$ respectively, relative to a weighted mean zero. The corresponding values for the same ligands in a tetrahedral arrangement (where the doublet *e* level now has the lowest energy) will therefore be

$$e = -\tfrac{3}{5} \cdot \tfrac{4}{9}\,\Delta = -0\cdot27\,\Delta$$

and

$$t_2 = +\tfrac{2}{5} \cdot \tfrac{4}{9}\,\Delta = +0\cdot18\,\Delta$$

As we saw earlier the extent to which the ligands stabilize the *d* orbitals is called the 'crystal field stabilization energy'—abbreviated C.F.S.E.

The splitting produced by a square-planar arrangement of ligands is shown in *Figure 12.17(b)*. The $d_{x^2-y^2}$ level is the least stable and is of much higher energy than the others. The d_{xy} orbital has the next highest energy, since the axes of its lobes all lie in the plane of the ligands. The d_{xz} and d_{yz} orbitals are degenerate, since they must be influenced by the ligand field to the same extent, but the experts differ on the energy of these orbitals relative to that of the d_{z^2} orbital. We shall use the scheme described by Basolo and Pearson where the d_{z^2} orbital is given an energy higher than that of d_{xz}, d_{yz}. This may be justified on the ground that the 'collar' of charge in the xy plane of the d_{z^2} orbital (*see* page 45) gives a greater repulsion with the ligand field.

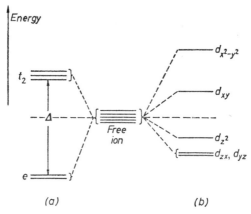

Figure 12.17. d-level splitting in (*a*) tetrahedral ligand field; (*b*) square-planar ligand field (not to scale)

The principal factors that determine whether a transition metal ion will form tetrahedral or square-planar complexes are *(a)* the crystal field stabilization energy, and *(b)* mutual repulsion between the ligands—which, in turn, will depend on the 'bulk' of the ligand and on its electronegativity. We shall first consider the C.F.S.E. values for tetrahedral complexes of metal ions with configurations from d^1 to d^{10}. The first electron will occupy the d orbital of lowest energy—*i.e.* one of the e orbitals, and in doing so will stabilize this level by $0.27\ \Delta$ energy units. The second d electron enters the second level of the doublet, and the C.F.S.E. value goes up to

$$2 \times 0.27\ \Delta = 0.54\ \Delta$$

There are two alternatives for the third electron; it may go either into the e levels, in which case it must 'pair' with an electron already in residence, to give a total C.F.S.E. of $0.81\ \Delta$, or it may go into one of the higher t_2 levels, giving a state with three unpaired electrons and a C.F.S.E. value of $(0.54 - 0.18)\ \Delta = 0.36\ \Delta$. There is, therefore, a competition between the additional crystal field stabilization produced by 'spin-pairing', and the operation of Hund's rules, similar to that discussed earlier for octahedral complexes. The choice may be determined by the strength of the ligand field; if the field is weak the C.F.S.E. value is small, and the third electron goes into the t_2 orbital giving a 'spin-free' arrangement, whereas strong ligand fields favour 'spin-pairing' in the e level. The argument outlined here can be extended to other d electron configurations ($d^4 \to d^{10}$) and to square-planar 4-co-ordination. The relevant C.F.S.E. values are shown in *Table 12.10*.

Table 12.10. Crystal Field Stabilization Energies (in Δ units) for Tetrahedral and Square-planar Complexes

Number of d electrons	Weak field			Strong field		
	Tetrahedral	*Square-planar*	*Difference (square-planar)- (tetrahedral)*	*Tetrahedral*	*Square-* planar	*Difference (square-planar)- (tetrahedral)*
0	0	0	0	0	0	0
1	0·27	0·51	0·24	0·27	0·51	0·24
2	0·54	1·02	0·48	0·54	1·02	0·48
3	0·36	1·45	1·09	0·81	1·45	0·64
4	0·18	1·22	1·04	1·08	1·96	0·88
5	0	0	0	0·90	2·47	1·57
6	0·27	0·51	0·24	0·72	2·90	2·18
7	0·54	1·02	0·48	0·54	2·67	2·13
8	0·36	1·45	1·09	0·36	2·44	2·12
9	0·18	1·22	1·04	0·18	1·22	1·04
10	0	0	0	0	0	0

* Assuming that even in a strong field the d_{z^2} and the d_{xz} and d_{yz} levels are sufficiently close for one electron to go into each orbital before pairing occurs. Slightly different values will be obtained if the electrons are allowed to pair up in the d_{xz} and d_{yz} orbitals before entering the d_{z^2} orbital.

Table 12.10 shows clearly that if the C.F.S.E. were the only factor to be considered, then all 4-co-ordinate complexes would be square-planar, except the d^0, d^5 (spin-free) and d^{10} ones, where the C.F.S.E. is zero. However, the mutual repulsion between the ligands must also be taken into account, and this repulsion will be significant when the ligands are either electronegative or bulky. This steric factor will be particularly important when the field is weak and the C.F.S.E. correspondingly small, but repulsion will be rather less important in complexes formed by transition elements of the second and third rows where Δ is larger (compare tetrahedral $[NiCl_4]^{2-}$ and square-planar $[PtCl_4]^{2-}$). Tetrahedral complexes are known for all divalent elements of the first transition series, although up till now none with a spin-paired configuration have been characterized. Such spin-pairing is only likely to occur for large values of Δ (produced either by ligands giving a strong field or with metals of the second and third transition series), and we shall see that when Δ is large we get square-planar rather than tetrahedral complexes. Another factor to note is the tendency for compounds (especially of the heavier elements) to polymerize.

We now discuss some examples of 4-co-ordination for metal ions of different d electron configurations.

d^0, d^5 (spin-free) and d^{10}

Here C.F.S.E. is zero, because the orbitals are either empty, all singly-filled or all doubly-filled (e.g. if each orbital contains one electron, the C.F.S.E. value is

$$[2 \times (- 0.27 \, \Delta) + 3 \times (+ 0.18 \, \Delta)] = 0.)$$

In each case the ligands will take up the tetrahedral position which minimizes the repulsion between them. Tetrahedral arrangements are thus found in $TiCl_4$, $[FeCl_4]^-$ and $[ZnX_4]^{2-}$, although the latter two are slightly distorted because of the influence of the cations in the crystal lattice.

d^1, d^6

With either one or six d electrons, the C.F.S.E. is very small, so we should expect the repulsion factor to be the dominating one, and indeed a tetrahedral configuration is found for the d^1 compound VCl_4; NbI_4, however, is polymeric (DAHL and WAMPLER[43]) with octahedrally co-ordinated niobium atoms forming long chains through shared edges. No d^6 4-co-ordinate compounds appear to be known.

d^2, d^7

The C.F.S.E. is still rather small, and it is unlikely to be the deciding factor for first row transition elements. No d^2 complex has been properly characterized up till now, but the tetrahedral configuration is certainly present in many divalent cobalt complexes (d^7 configuration). Thus the tetrahedral arrangement is found both in anions, $[CoX_4]^{2-}$ ($X = $ Cl, Br, I, NCS), and in neutral compounds of the type CoX_2v2L (with $X = $ halogen, and $L = $ pyridine, p-toluidine, and triphenylphosphine oxide). HOLM and COTTON[44] have recently discussed these Co(II) complexes in some detail, and point out that the experimental values obtained for magnetic susceptibilities (4.4–5.0 B.M.) are appreciably higher than the value (3.87 B.M.) given by the spin-only formula for three unpaired electrons. The increase is attributed to orbital contribution.

A square-planar configuration is observed (FORRESTER et al.[45]) for divalent cobalt in the maleonitrile dithiolate complex $(NBu_4^n)_2$ $[CoS_4C_8N_4]$, and this has now been shown to be a low spin complex (DAVISON et al.[46]).

Two apparently 4-co-ordinate cobalt(II) complexes, $Co(acac)_2$ (COTTON and ELDER[47]) and $Co(OPMe_3)_2(NO_3)_2$ (COTTON and SOLDERBERG[48]) both contain octahedrally co-ordinated Co atoms, the first compound being tetrameric with bridging acetylacetone groups, and the second containing bidentate nitrate groups.

d^3, d^4

No 4-co-ordinate complexes appear to have been characterized for metals of the first transition series with these configurations, but if they were to be prepared we might expect them to be planar in view of the large C.F.S.E.

Potentially four-co-ordinate compounds of technetium(IV) (d^3) and rhenium(III) (d^4) turn out to be polymeric. Thus $TcCl_4$ has a structure (ELDER and PENFOLD[49]) analogous to that of NbI_4, with octahedral technetium atoms bridging through shared edges to give endless chains. The rhenium(III) complexes $[ReCl_4]^-$ (ROBINSON et al.,[50] BERTRAND et al.[51]) and $ReCl_3(PEt_2Ph)$ (COTTON and MAGUE[52]) are both trimeric.

The rhenium atoms are seven-co-ordinate, being bonded in the plane (xy) to two other Re atoms, two bridging Cl atoms, and a ligand L, and to two terminal atoms above and below the plane.

d^8

On grounds of C.F.S.E. this configuration should always give rise to square-planar complexes. The complexes of the heavier metals, palladium and platinum, are indeed planar, irrespective of the nature of the ligands, and, furthermore, they are spin-paired and diamagnetic. Amongst the complexes of divalent palladium and platinum that have been shown to have a square-planar structure by x-ray methods may be mentioned K_2PdCl_4, K_2PtCl_4, and $(NEt_4)_2 [Pt_2Br_4]$ (STEPHENSON[53]); the latter complex has the expected bridged arrangement:

$$\left[\begin{array}{ccc} Br & Br & Br \\ & \diagdown \diagup \searrow \diagup & \\ Pt & & Pt \\ & \diagup \nwarrow \diagup \diagdown & \\ Br & Br & Br \end{array} \right]^{2-}$$

Palladium(II) derivatives of N-alkylsalicylaldimine have been shown (FRASSON et al.[54]) to have the *trans* square planar structure:

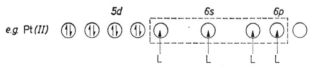

The ligand-field approach leaves the $d_{x^2-y^2}$ orbital empty, and it is worth remembering that in the valence-bond description of such complexes it is this orbital that is mixed with the s, and the p_x and p_y orbitals to give the dsp^2 hybrids which accept electrons from the four ligands.

e.g. Pt(II)   5d ⇅ ⇅ ⇅ ⇅   6s ↑ ↑   6p ↑ ↑ ○
                                      L   L   L L

The position with 4-co-ordinate complexes of nickel(II) is not so clear cut. Before the development of ligand-field theory, nickel (II) complexes were divided in two classes (a) diamagnetic complexes, which were normally yellow-red in colour with a strong peak around 4000 Å in their absorption spectra, and (*b*) para-magnetic complexes, which were usually blue-green in colour with no such 4000 Å peak. The diamagnetic complexes were considered to be square-planar, using dsp^2 orbitals, and the para-magnetic complexes were assumed to be tetrahedral (sp^3 orbitals).

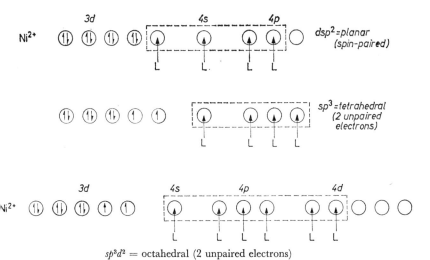

sp^3d^2 = octahedral (2 unpaired electrons)

The magnetic properties were often taken as diagnostic of the configuration.

This magnetic criterion for tetrahedral/square-planar complexes is misleading, however, because it is by no means essential for paramagnetic complexes to be tetrahedral. Thus magnetic moments appropriate to two unpaired electrons are also to be expected for any octahedral nickel complexes, and the acetyl-acetone complex, Ni($acac$)$_2$, has been shown by x-ray studies (BULLEN *et al*.[55]) to be trimeric, with octahedral nickel. The valence-bond description is then one involving outer $4s4p^34d^2$ orbitals.

The structures of quite a number of diamagnetic nickel(II) complexes have been determined by x-ray techniques and shown to be square-planar. These include *bis*-(ethylmethylglyoxime) nickel(II)[56], dibromo-*bis*-(triethylphosphine) nickel(II), and the *bis*-(maleonitriledithiolate)nickel(II) anion[57]; the first two are illustrated by formulae A and B and the last is the same as that shown for the analogous cobalt (II) complex on page 249.

A

B

When nickel is four-co-ordinate, the ligand-field approach indicates that we should expect diamagnetic square-planar structures unless the ligands give only a weak field, or are very electronegative, or bulky. This is supported by recent x-ray work (GARTON et al.[58]) that has shown $NiCl_2,2PPh_3$ to have a distorted tetrahedral structure, and by the considerable evidence (cf. GILL and NYHOLM[59], GRUEN[60]) showing that the tetrahalogeno anions $[NiX_4]^{2-}$, $X =$ Cl, Br, and I, are tetrahedral as salts of large cations and in some fused melts.

It is, however, very dangerous to make too many predictions about the stereochemistry of four-co-ordinate nickel(II) complexes, since the factors determining whether the tetrahedral or the square-planar arrangement is the more stable are very delicately balanced. X-ray structural work is vital. Thus a recent study of $NiBr_2$, $2PPh_2$Benzyl, which is a green paramagnetic complex ($\mu = 2.7$ B.M.), shows the unit cell to contain two tetrahedral and one square-planar nickel atom. Work on the bis-(N-alkylsalicyl-aldiminato)nickel(II) complexes provides further evidence for the delicate balance. Thus the N-methyl compound is square-planar (FRASSON et al.[61]) but the N-isopropyl complex has a distorted tetrahedral structure (FOX et al.[62]). Furthermore, CHAKRAVORTY et al.[63]) have carried out experiments in which the N-alkyl group contains a site capable of further co-ordination (e.g. CH_3 CH : CH_2COCH_3) and have shown that in chloroform solution there is an equilibrium betweens quare-planar, tetrahedral, and octahedral forms, with the latter predominating at low temperatures.

d^9

Bivalent copper resembles nickel in that although square-planar complexes are almost always formed, tetrahedral complexes can be prepared with suitable ligands. Of the many known square-planar complexes we quote just two, $Cu(acac)_2$ and gaseous $Cu(NO_3)_2$ (LAVILLA and BAUER[64]) :

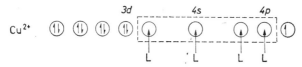

The valence-bond explanation of the square-planar configuration leaves the unpaired electron in an 'exposed' $4p$ orbital.

$$Cu^{2+} \quad \begin{array}{cccc} 3d & & 4s & 4p \end{array}$$

This electron should be easily lost, giving Cu(III) complexes, and since such oxidation does not occur, it is necessary to place the unpaired electron in a $3d$ orbital and use the $4d_{x^2-y^2}$ orbital to form the dsp^2 hybrids. In the ligand-field description, this unpaired electron is also placed in the $3d_{x^2-y^2}$ orbital.

As we saw earlier in this chapter (*see* page 232), a closer examination of many 'square-planar' complexes of divalent copper shows that they can usually be regarded as octahedral with two long (z axis) and four short bonds. This was discussed on the basis of the uneven filling of the e_g levels, with two electrons in the d_{z^2} orbital repelling the two z axis ligands. A somewhat more sophisticated interpretation invoked the Jahn-Teller theorem and gave a splitting of the e_g levels (*see Figure 12.11*). Any asymmetry in the z axis also results in a splitting of the t_{2g} triplet, because an electron in the d_{xy} orbital experiences a bigger ligand field than one in either the d_{xz} or d_{yz} orbitals. This is shown in *Figure 12.18*, which indicates that in the limit, when the two z axis ligands are removed so far away as to exert no influence, we get the ligand-field diagram for square-planar complexes.

We may illustrate the types of 4/2 'octahedral' copper (II) complexes by further reference to the complex chlorides $MCuCl_3$[36, 37] which we described earlier as containing planar $[Cu_2Cl_6]^{2-}$ units. In fact, these units are stacked so that the Cu atom in one unit comes between Cl atoms of the units immediately above and below. In this way each copper forms four short (~2·3 Å) and two long (3·11 Å) bonds. Similarly, the Cu atoms in the *trans*-planar complex CuCl$_2$,2*py* are linked through two longer Cu—Cl bonds (3·05 Å), and in the salicylaldoximato complexes (JARSKI and LINGAFELTER,[65] ORIOLI *et al.*[66]), the square-planar units are linked through two longer Cu—O bonds.

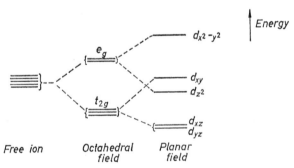

The only tetrahedral complexes of bivalent copper are the tetra-halogeno ions [CuCl$_4$]$^{2-}$ and [CuBr$_4$]$^{2-}$, where it seems that the repulsion between the electronegative halogen atoms and the relative small ligand field is sufficient to overcome the loss in C.F.S.E. It must be noted that there are only five electrons in the three t_2 levels (which point towards the ligands), so that the ligands will tend to be attracted towards this 'hole' in the electron charge density. In [CuBr$_4$]$^{2-}$, for instance (MOROSIN and LINGA-FELTER[67]), the tetrahedron is flattened such that the BrCuBr angles range from 100°–130° ; a similar distortion is found in [CuCl$_4$]$^{2-}$.

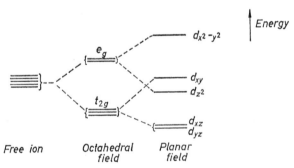

Figure 12.18. *d*-level splitting for octahedral and square-planar fields

The absorption spectrum of the [CuCl$_4$]$^{2-}$ ion in solution (FERGUSON[68]) is the same as that of the solid Cs$_2$CuCl$_4$, and the ion in solution is accordingly believed to be tetrahedral. Recent x-ray work (WILLETT[69]) on the solid ammonium salt shows it to contain planar [CuCl$_4$]$^{2-}$ ions (Cu—Cl = 2·30 Å), however, linked through two longer Cu—Cl bonds (2·79 Å), and it has been suggested that in the solid salts of $[CuCl_4]^{2-}$ either structure may be present, the planar salts being light yellow and the tetrahedral ones orange.

CO-ORDINATION NUMBER 5

Compounds of co-ordination number 5 are rather uncommon, and many apparently 5-co-ordinate compounds are really mixtures or polymers. We saw in Chapter 11 that solid PCl_5 contains a mixture of $[PCl_4]^+$ and $[PCl]_6^-$ ions, and it is known that Cs_3CoCl_3 contains $[CoCl_4]^{2-}$ and Cl^- ions rather than $[CoCl_5]^{3-}$. The solid pentachlorides of niobium, tantalum, and molybdenum are dimeric with slightly distorted octahedral structures.

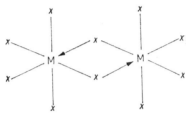

There are two basic configurations that can be adopted by the complex compounds of co-ordination number 5, the trigonal-bipyramidal and the square-pyramidal (cf. Figure 12.19 (a) and (b)).

Since the original publication by Kimball, there has been quite a lot of discussion about the hybrid orbitals used in these two configurations, particularly about the combination dsp^3, which appears to be used for both bonding arrangements. This is understandable if we suppose that the d_{z^2} orbital is used for trigonal-bipyramidal bonds and the $d_{x^2-y^2}$ for square-pyramidal bonds. It may give a clearer picture if we think of trigonal-bipyramidal orbitals as two different sets of hybrids, sp^2 (giving the three trigonal-

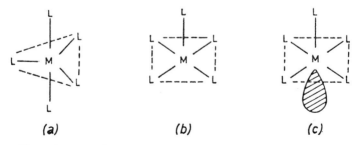

(a)   (b)   (c)

Figure 12.19. Configurations for complexes of co-ordination number 5: (a) trigonal-bipyramidal; (b) square-pyramidal; (c) octahedral

planar bonds) $+ d_{z^2}p_z$ (giving the two linear axial bonds) ; in the same way the square-pyramidal hybrids can be viewed as $d_{x^2-y^2}sp^2$ (giving four square-planar bonds) $+ p_z$ (giving the apical bond). We shall see that the square-pyramidal structure almost always arises when a lone pair of electrons is present, so that we can consider the structure as octahedral with the lone pair taking up one position (*cf. Figure 12.19(c)*).

In predicting which of the structures is the more likely we must allow for a number of factors (GILLESPIE[70]) :

(i)   steric requirements of ligands,

(ii) mutual repulsion of bonding pairs and lone pairs of electrons,

(iii) mutual repulsion of bonding pairs and any d-shell electrons,

(iv) C.F.S.E.

If we limit our discussion for the moment to compounds of the non-transition elements, then only the first two factors need be considered. On the basis of (ii) we expect to get the trigonal-bipyramidal structure for all MX_5 compounds, except when there is a lone pair of electrons ; in the latter case we should get the 'octahedral' structure shown in *Figure 12.19 (c)*. Compounds of both these types were discussed in Chapter 11, page 187. The only additional non-transition element compound worth mentioning is SOF_4, which we saw (page 198) has a trigonal-bipyramidal structure with the $S = O$ in an equatorial position.

When we come to consider known 5-co-ordinate compounds of the transition elements then the position is much more complicated. *Table 12.11* lists the C.F.S.E. for the two configurations, from which

Table 12.11. C.F.S.E. for 5-Co-ordinate Complexes
C.F.S.E. (in Δ units)

No. of d electrons	Trigonal-bipyramidal	Square pyramid	Difference (S.P. − T.B.P.)
0	0	0	0
1	0·27	0·46	0·17
2	0·54	0·92	0·38
3	0·62	1·01	0·39
4	0·70	0·92	0·22
5	0·97	1·38	0·41
6	1·24	1·84	0·60
7	1·32	1·93	0·61
8	1·40	1·84	0·44
9	0·70	0·92	0·22
10	0	0	0

it can be seen that the square-pyramidal structure is always the most stable but not enormously so.

The figures given in *Table 12.11* are based on the splitting diagrams shown in *Figure 12.20* and the values quoted in BASOLO and PEARSON.[4] It is assumed in each case that the first four levels are filled singly, then doubly, with the highest energy level only being occupied for d^9 and d^{10} complexes.

Table 12.12. Examples of 5-co-ordinate Complexes*

No. of d electrons	Trigonal-bipyramidal	Square-pyramidal
0	(i) MX_5 (*e.g.* $NbCl_5$ vapour) (ii) $TiCl_4,NMe_3$ (71, 72)	—
1	$MoCl_5$ vapour	$VO(acac)_2$ (73) $[VO(SCN)_4]^{2-}$ (74)
2	$VX_3,2L$ (*e.g.* $VCl_3,2NMe_3$) (75, 72)	—
3	—	$[Re_2Cl_8]^{2-}$ (76)
6	—	$RuCl_2,3PPh_3$ (77)
7	—	$Co(S_2CNMe_2)_xNO$
8	(i) $Fe(CO)_5$ (ii) $[Co(CNMe)_5]^+$ (79) (iii) $[M(QAS)X]^+$ (M = Ni and Pt) (80, 81)	$NiBr_2, triars$ (78)
9	(i) $[CuCl_5]^{3-}$ (82) (ii) $Cu(terpyr)Cl_2$(83) (iii) $[Cu(bipyr)_2I]^+$ (84)	—
10	$Zn(acac)_2\ H_2O$ (85)	

* References given in parentheses

Table 12.12 lists the 5-co-ordinate complexes of various d^n configurations that have been characterized as either trigonal-bipyramidal or square-pyramidal.

It may be noted that the trigonal-bipyramidal structure always arises when the five ligands are equivalent, irrespective of the nature of the ligand or the number of d electrons. As we shall see later in this chapter, we cannot really talk about the number of d electrons in $Fe(CO)_5$ because there is considerable delocalization of these electrons through d_π—p_π bonding. There appears to be similar π bonding in the $[Co(CNMe)_5]^+$ cation, since the Co—C bond length of 1·87 Å is much less than anticipated for a single bond. Some of the other trigonal-bipyramidal complexes justify further comment. Consideration of far infra-red spectra (which gives information about the arrangement of metal-chlorine bonds) shows that in both $TiCl_4 \cdot NMe_3$ and $VCl_3 \cdot 2NMe_3$ the bulky NMe_3 groups occupy axial positions, and the *trans* configuration of the vanadium complex is confirmed by its very low dipole moment. The nickel and platinum complexes formed with the quadridentate arsine ligand (QAS) have an 'octopus' type of structure with the four arsenic atoms occupying one axial and the three equatorial positions :

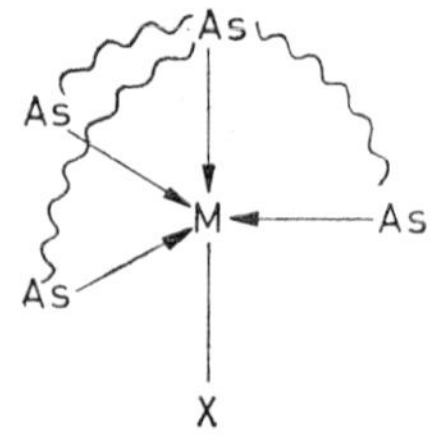

The copper(II) complexes $[CuCl_5]^{3-}$, $CuCl_2 \cdot terpyr$, and $[CuI(bipyr)_2]^+$ form an interesting series in which the trigonal-bipyramidal configuration is retained. The axial positions in the latter two complexes are occupied by nitrogen atoms, presumably because the 90° NCuN angles cause less strain in the chelate rings. In the acetylacetone complex of zinc, the structure is considerably distorted from the ideal trigonal-bipyramid because of the requirements of the bidentate *acac* groups.

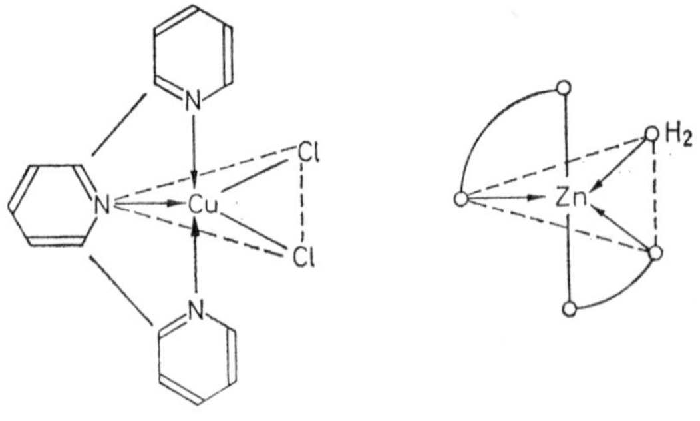

While the trigonal-bipyramidal structures are accounted for quite readily in terms in electron-pair repulsions and steric requirements of the ligands, explanations are not so convincing for the square-pyramidal molecules. Thus $VO(acac)_2$ (I) and other diketone complexes have the VO bond along the z axis with the vanadium atom significantly above the plane of the four oxygen atoms from the chelating ligands, although we might have expected a trigonal-bipyramidal molecule with the VO bond in one of the equatorial positions. Gillespie suggests that the single d electron might occupy the sixth position for an octahedral arrangement, but the same argument should apply to $MoCl_5$ and this is trigonal-bi-pyramidal!. In solution, the complex is octahedral because a molecule of solvent is invariably co-ordinated. Similarly, the $[VO(SCN)_4]^{2-}$ anion has a co-ordinated water molecule occupying the sixth position. Readers interested in complexes of this type are referred to a paper by BALLHAUSEN and GRAY[86] which deals with the problem (especially of spectra) in some detail.

| I | II | III |

The $[Re_2Cl_8]^{2-}$ ion (II) is very interesting in that it consists of two $ReCl_4$ units linked by an Re—Re bond, the Re—Cl bonds being bent away from the Re—Re link. What is unexpected is that the two $ReCl_4$ units are eclipsed rather than staggered. The ruthenium complex, $RuCl_2 \cdot 3PPh_3$ again might have been predicted to have a trigonal-bipyramidal structure, with the three phosphine groups in equatorial positions, rather than that shown in III. Moreover the compound is diamagnetic, which means that two electrons pair up in the d_{xy} orbital rather than going singly into this and the d_{z^2} orbital. The authors (LA PLACA and IBERS[77]) point out that the structure is probably 5-co-ordinate only because the bulky phenyl groups block the sixth position. We should, perhaps, then expect the difference between the d_{xy} and d_{z^2} orbitals to be much greater than shown in *Figure 12.20* (the diagram being midway between this and an octahedral one), so that diamagnetism would then be reasonable.

The nickel(II) compound formed by the triarsine ligand is by no means a regular square pyramid, and it would seem that the steric requirements of the tridentate ligand may be the determining factor.

We can perhaps summarize 5-co-ordination in transition metal complexes by saying that while the trigonal-bipyramidal structure

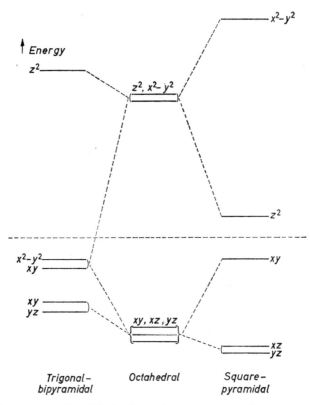

Figure 12.20. Ligand-field splitting diagram for 5-co-ordinate complexes

is the commonest, the square-pyramidal one arises occasionally. A great deal more experimental work is necessary before all the factors can be evaluated.

CO-ORDINATION NUMBER 7 [87]

Compounds with this co-ordination number are rare, and many of the apparently 7-co-ordinate substances are mixtures. $(NH_4)_3SiF_7$, for instance, contains a mixture of $[SiF_6]^{2-}$ and F^- ions and no $[SiF_7]^{3-}$ ions. Even where 7-co-ordination has been established there is often no information about the stereochemistry. Thus quite a number of substituted carbonyl derivatives of molybdenum and tungsten, *e.g.* $[Et_4N][M(CO)_4Br_3]$ and $Mo(CO)_2(diars)_2Br$ (GANORKAR and STIDDARD[88]; NIGAM *et al.*[89]) have been established as 7-co-ordinate, and the 1 : 1 diarsine adducts of niobium(V) and tantalum(V) chloride and bromide, $MX_5, diars$, are monomeric in solution (CLARK et al.[90]).

In Chapter 11, we discussed iodine(VII) fluoride, and considered that it probably had a pentagonal-bipyramidal distribution of fluorine atoms about the iodine; the vibrational spectrum of rhenium(VII) fluoride has been studied recently and interpreted on the basis of the same structure (CLAASSEN and SELIG[91]). X-ray studies show the same arrangement in several complex fluoride ions, e.g. $[ZrF_7]^{3-}$, $[UF_7]^{3-}$, and $[UO_2F_5]^{3-}$.

We saw earlier in this chapter (page 249) that in $CsReCl_4$ the anion was really $[Re_3Cl_{12}]^{3-}$ in which each rhenium was linked to two other rhenium and five chlorine atoms, thus achieving a distorted pentagonal-bipyramidal structure.

Two other configurations are predicted (see Table 12.2, page 215): an octahedron with an extra atom at the centre of one face and the remaining six bonds slightly distorted (face-centred octahedron) is one possibility; a trigonal prism with the extra atom at the centre of one square face (face-centred trigonal prism) is the other. No definite examples of the face-centred octahedral structure are known, although $[ZrF_7]^{3-}$ was originally considered to have this arrangement. The face-centred trigonal-prism structure is reported for the ions $[NbF_7]^{2-}$ and $[TaF_7]^{2-}$, which is rather surprising since they have analogous electronic configurations to $[ZrF_7]^{3-}$, and might be expected to adopt a similar structure.

CO-ORDINATION NUMBER 8

Generally, it has been considered that eight co-ordination could be expected only for the heavier elements because those of the first transition series would be too small; furthermore, the larger energy gap between the $3d$, $4s$, and $4p$ orbitals as compared with the $4d$, $5s$, and $5p$, and $5d$, $6s$, and $6p$ orbitals reduces the possible hybridization.

Thus, although both molybdenum and tungsten form octacyanide complexes, $K_4M(CN)_8$, the analogous chromium compound has not been made. As a further illustration, we mention the complex fluoride ion $[TaF_8]^{3-}$; neither vanadium nor niobium give analogous ions, although a seven co-ordinate oxyfluoride niobium ion $[NbOF_6]^{3-}$ can be obtained. Quite recently, however,

two eight co-ordinate titanium compounds have been characterized, TiCl$_4$,2*diars* (CLARK *et al.*[92]) and Ti(NO$_3$)$_4$ (ADDISON *et
al.*[93]), so it would appear that others might well be prepared. The
analogous diarsine complexes of MCl$_4$ (M = Zr, Hf, V, and Nb),
MBr$_4$ (M = Ti, Zr, Hf, Nb), and NbI$_4$ have also been characterized (*cf.* CLARK *et al.*[94]).

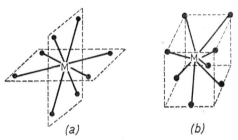

Figure 12.21.   Configurations for complexes of co-ordination number 8:
(a) dodecahedral; (b) square-antiprismatic

The two commonest configurations found for this co-ordination
number, dodecahedral and square antiprismatic, are illustrated in
Figure 12.21. The valence-bond approach uses d^4sp^3 hybrids for
the dodecahedral and d^5p^3 hybrids for the square antiprismatic
structure. In recent years the general theory of eight co-ordination
has been discussed by several sets of workers (HOARD and SILVER-
TON[95]; RANDIĆ[96]; KEPERT[97]). RANDIĆ for instance has discussed
the ligand-field splittings, and KEPERT has made calculations of
ligand-ligand repulsion energies and found that although on this
basis the dodecahedral and square antiprismatic arrangements are
considerably more stable than any other, the most stable configurations nevertheless involve bond angle distortions of several degrees
from the ideal. *Table 12.13* lists various eight co-ordinate complexes that have been reported for the two configurations.

As the examples in the table indicate, there is no apparent reason
why one stereochemistry is preferred to the other for a particular
complex; thus both configurations are observed with both monodentate and bidentate ligands. Moreover, whereas thorium is
eight co-ordinate in its tetrahalides (through bridging halogen
atoms), the arrangement is dodecahedral for the chloride and square
antiprismatic for the iodide (ZALKIN *et al.*[101]).

The only other stereochemistry found experimentally is the
hexagonal-bipyramidal one, and this is for the rather special case
of uranyl complexes UO$_2$(NO$_3$)$_2$L$_2$, (FLEMING and LYNTON[102]),

Table 12.13. Complexes of Co-ordination Number 8

Configuration	Compounds	Reference
Dodecahedral	$K_4Mo(CN)_8$	(1)
	$[Zr(C_2O_4)_4]^{4-}$	(95)
	$TiCl_4,2diars$	(92)
	$Ti(NO_3)_4$	(93)
	$[Co(NO_3)_4]^{2-}$	(98)
Square antiprismatic	$M(acac)_4$ (for M = Zr, Th, Ce and U)	(99)
	$Ce(PhCOCHCOPh)_4$	(100)
	$[TaF_8]^{3-}$	(1)

where $L = H_2O$ and $OP(OEt)_3$. The two nitrate groups act as bidentate ligands, and together with the two ligands (L) give six planar (xy) bonds (I); the two $U = O$ bonds form along the z axis (II).

I   II

From a stereochemical point of view these nitrogroups can be regarded alternatively as monodentate with the metal-ligand direction bisecting the ONO angle, so that the uranium complexes can then be looked on as octahedral. On this basis titanium (IV) nitrate (cf. Table 12.13) is tetrahedral, and it is interesting to note that the absorption spectrum of the dodecahedral cobalt anion, $[Co(NO_3)_4]^{2-}$, is typical of tetrahedral cobalt(II) species (COTTON and BERGMAN[98]).

CARBONYLS, NITROSYL CARBONYLS, AND CARBONYL HYDRIDES[103]

The carbonyls (and the related nitrosyl carbonyls and carbonyl hydrides) are best discussed as a group, rather than under the individual co-ordination numbers. The carbonyls, listed in Table 12.14, are diamagnetic, with inert gas configurations (except $V(CO)_6$), and are unique in that the metal atoms are in zero valency states.

In all the carbonyls which have been studied in detail, the CO groups are linked to the metal through the carbon atom with

Table 12.14. Carbonyls

Group V	Group VI	Group VII	Group VIII		
$V(CO)_6$	$Cr(CO)_6$	$[Mn(CO)_5]_2$	$Fe(CO)_5$ $Fe_2(CO)_9$ $[Fe(CO)_4]_3$	$[Co(CO)_4]_2$ $[Co(CO)_3]_4$	$Ni(CO)_4$
	$Mo(CO)_6$	$[Tc(CO)_5]_2$	$Ru(CO)_5$ $[Ru(CO)_4]_3$	$[Rh(CO)_4]_2$ $[Rh(CO)_3]_n$ $Rh_6(CO)_{16}$	
	$W(CO)_6$	$[Re(CO)_5]_2$	$Os(CO)_5$ $[Os(CO)_4]_3$	$[Ir(CO)_4]_2$ $[Ir(CO)_3]_n$	

collinear M—C and C—O bonds. In nickel carbonyl, the four CO groups are arranged tetrahedrally around the nickel atom, and we can write the molecule as

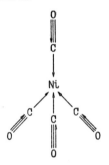

in which the nickel atom has the configuration

The Ni ← C σ bonds are described in valence-bond language as resulting from the overlap of vacant sp^3 hybrid orbitals of Ni with doubly-filled σ orbitals of the four C atoms. These, and other appropriate atomic orbitals for one Ni ← C ═ O grouping, are illustrated in *Figure 12.22*.

It is significant that carbon monoxide does not form similar bonds with good electron acceptors such as boron trifluoride, and it seems that some additional d_π—p_π bonding must strengthen the Ni—C linkages (*cf.* olefin complexes of Pt, page 276). Further-

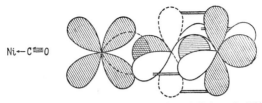

$Ni \leftarrow C \equiv O$

Figure 12.22. Orbitals for one Ni $\leftarrow$ C $\leqq$ O linkage in Ni(CO)$_4$

more, if we rewrite the formula $Ni\leftarrow(C\leqq O)_4$ as $Ni^{4-}—(C\equiv O^+)_4$, by putting in the formal charges, we notice a build-up of negative charge on the nickel atom. This charge can be partially removed by distortion of the Ni—C σ molecular orbital, carbon being more electronegative than nickel, but most of it will be removed by back donation of charge from the doubly-filled d orbitals of the nickel atom to the vacant π orbital on the carbon atom. This $d_\pi—p_\pi$ bonding also accounts for the Ni—C bond length (1·84 Å), which is appreciably shorter than that calculated for the single bond (2·00 Å); the M—C bond lengths in the other metal carbonyls are also significantly shorter than the single bond values.

If we look again at *Figure 12.22* we see that the $d_\pi—p_\pi$ bonding can be delocalized to include the oxygen atom, since the latter has a suitable p_π orbital. For the moment let us take an oversimplified picture and assume that the metal d orbital overlaps with the p_π orbital of just one carbon atom. Then we can apply the simple molecular-orbital procedure (*cf.* CO_2 in Chapter 11) and combine the three atomic π orbitals (metal d, carbon p, and oxygen p) to give three delocalized π molecular orbitals, one bonding, one non-bonding, and one antibonding; the four electrons are placed in the first two of these molecular orbitals. As we saw earlier in this chapter, there is a maximum of only three π bonds for octahedral complexes, so that in the hexacarbonyls the maximum M—C π bond order is 0·5. For Ni(CO)$_4$ and Fe(CO)$_5$ $d_\pi—p_\pi$ bonding is also present but a discussion is complicated because all the M—C bonds are not directed along the Cartesian axes (x, y and z).

Recently x-ray[104] and electron-diffraction[105] measurements have confirmed the earlier view that iron pentacarbonyl has a trigonal-pyramidal structure; Fe(O) may take the configuration

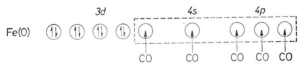

265

and use the d_{z^2} orbital to form dsp^3 hybrid orbitals which accommodate electrons donated by the carbon monoxide groups.

Chromium, molybdenum, tungsten and vanadium hexacarbonyls all have an octahedral structure with d^2sp^3 bonding. The vanadium compound, a dark green solid (CALDERAZZO et al.[106]), is particularly interesting since it has an unpaired electron; all the other carbonyls are diamagnetic, and it would seem that vanadium hexacarbonyl is unable to dimerize—possibly for steric reasons.

The structures of some of the polynuclear carbonyls are now known with reasonable certainty, but there are others where the structure is still a matter for speculation. The structures of the dimeric manganese, technetium and rhenium carbonyls have recently been examined (DAHL et al.[107]), and found to contain octahedrally co-ordinated metal atoms; each metal atom is attached to five CO groups, and is also directly bonded to the other metal atom, the linked octahedral units being staggered from the eclipsed position by an angle of about 45°.

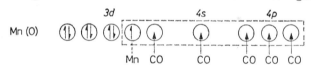

The manganese atom assumes the octahedral configuration

Mn (0)

with the d orbitals containing the unpaired electron overlapping to give a Mn—Mn bond and a diamagnetic molecule. The metal–metal bond in both these carbonyls is rather longer than that expected for a single bond however, possibly because of repulsions between neighbouring CO groups belonging to different metal atoms.

This linkage of the two halves of the carbonyl molecule by a single metal–metal bond is only found in the manganese, technetium and rhenium compounds; in the other binuclear carbonyls the metal atoms are additionally linked by bridging carbonyl groups. Thus, in $Fe_2(CO)_9$, (I) three carbonyl groups are bonded to each iron atom in the usual way, and the remaining three carbonyl groups form bridges between the iron atoms; an iron-iron link is also needed to account for the diamagnetism of the molecule, and the Fe—Fe distance (2·5 Å) is roughly that expected for a single

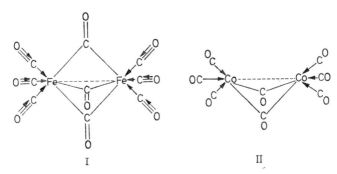

I   II

bond.   The octacarbonyl of cobalt, $Co_2(CO)_8$ has a closely related structure (SUMNER *et al.*[108]) with one of the bridging CO groups missing.

Besides the conclusive proof of x-ray crystal structure determination, additional evidence for bridging CO groups is provided by infra-red spectra.   Thus CO stretching frequencies for terminal carbonyl groups are found in the 1,900–2,050 cm^{-1} region, while those for bridging groups are found at lower frequencies (1,800–1,900 cm^{-1}).   Readers interested in the infra-red spectra of the metal carbonyls are referred to ABEL's review[103] where the limitations of structural applications are discussed.

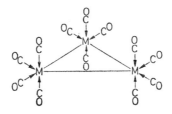

III

The structures of several other polynuclear carbonyls have been examined. The osmium and ruthenium dodecacarbonyls, $M_3(CO)_{12}$, are isomorphous (COREY and DAHL[109]), and contain the three metal atoms linked by metal–metal bonds in an equilateral triangle, there being no bridging carbonyl groups.

The structure of the analogous iron carbonyl $Fe_3(CO)_{12}$ has caused a lot of trouble and it has not yet been determined unambiguously. One early suggestion was structure IV, with a linear arrangement of iron atoms linked through bridging carbonyl groups, but this appeared to be ruled out by preliminary x-ray data (DAHL and

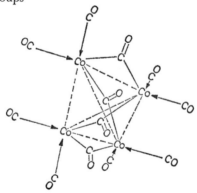

IV   **V**

Rundle[110]) which indicated a triangular arrangement for the three metal atoms (*cf.* $Os_3(CO)_{12}$).

Recently, the infra-red spectra (Dobson and Sheline[111]) has been interpreted on the basis of a linear structure with bridging CO groups, and Mössbauer spectra indicate (Herber *et al.*[112]) two different types of iron atom. An x-ray study of the closely related anion $[HFe_3(CO)_{11}]^-$ (Dahl and Blount[113]) provides the latest, and possibly the most valuable information. In the anion, structure V, the three iron atoms do form a triangle, but one pair is linked through both a CO group and an H atom. It is accordingly suggested that $Fe_3(CO)_{12}$ has a similar structure with a CO group replacing the H atom. The dodecarbonyl is then regarded as formed from $Fe_2(CO)_9$ by the insertion of a *cis* $Fe(CO)_4$ grouping instead of one of the bridging CO groups.

$Co_4(CO)_{12}$ has been examined further (Smith[114]) and shown to have structure VI in which there are eight terminal and four bridging carbonyl groups

The polynuclear rhodium carbonyl originally considered to be $Rh_4(CO)_{11}$ has now (Corey *et al.*[115]) been shown to be $Rh_6(CO)_{16}$ and to have a structure in which the six rhodium atoms are located at the corners of an octahedron with twelve terminal and four bridging carbonyl groups; the latter are located above four of the octahedral faces and apparently bridge three rhodium atoms.

In the nitrosyl carbonyls and carbonyl hydrides, $Mn(CO)(NO)_3$, $Mn(CO)_5H$, $Fe(CO)_2(NO)_2$, $Fe(CO)_4H_2$, $Co(CO)_3(NO)$ and $Co(CO)_4H$, the metal atom again achieves the inert gas configuration, giving compounds isoelectronic with $Ni(CO)_4$. Each H, CO, and NO may be taken as contributing one, two and three electrons respectively to the metal, thus producing a 'pseudo-nickel' atom. In the carbonyl hydrides of iron and cobalt the four CO groups are arranged tetrahedrally around the metal atoms; the hydrogen atoms appear to be attached directly to the metal atoms, although there seems to be some interaction with the nearby CO groups (EDGELL and GALLUP[116]). There is also a Mn—H bond in $Mn(CO)_5H$ (EDGELL et al.[117]).

SUBSTITUTED CARBONYL COMPOUNDS[118]

The CO groups in carbonyls can be displaced either by (a) typical ligands such as, e.g., phosphines, or by (b) π-donating groups such as *cyclo*pentadiene, benzene, olefins, etc.

(a) If a metal carbonyl reacts with π-bonding monodentate ligands such as triphenylphosphine, then one or more CO groups may be replaced by an equivalent number of ligands. If a bidentate ligand such as o-phenylene*bis*dimethylarsine (*diars*) is used, then each ligand molecule displaces two CO groups. *Table 12.15* summarizes some of the compounds that have been prepared from mononuclear carbonyls. These substituted carbonyl compounds are monomeric; however, if non-π-bonding ligands such as amines are used, the products may be ionic rather than simple monomers.

Several metal carbonyl derivatives studied recently are especially interesting because they contain isolated bonds between dissimilar metals. Thus $Ph_3PAuCo(CO)_4$ (KILBOURN et al.[119]) has structure A in which the cobalt atom is trigonal-bipyramidal with the Co—Au

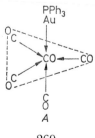

A

bond in an axial position. Mn—Au and Mn—Cu bonds are present in the analogous compounds $Ph_3PAuMn(CO)_5$ and *triars*CuMn $(CO)_5$.

(*b*) A range of compounds can be prepared by the reaction of metal carbonyls with π-electron donors such as *cyclo*pentadiene, benzene, various olefins and acetylenes (PAUSON[120]). We can illustrate this by reference to the *cyclo*pentadiene(Cp) compounds: $CpV(CO)_4$, $[CpCr(CO)_3]_2$, $CpMn(CO)_3$, $[CpFe(CO)_2]$, $CpCo(CO)_2$, $[CpNi(CO)]_2$.

In these compounds the metals again formally achieve the inert gas configuration, if we assume the *cyclo*pentadiene groups contribute all five of their π electrons, and that a metal–metal bond is

Table 12.15. Substituted Mononuclear Carbonyl Compounds

Carbonyl	Product with PR_3 etc.	Product with 'diars'
$Ni(CO)_4$	$Ni(CO)_3(PR_3)$ $Ni(CO)_2(PR_3)_2$ $Ni(CO)(POPh)_3$	— $Ni(CO)_2(diars)$ —
$Fe(CO)_5$	$Fe(CO)_4(PR_3)$ $Fe(CO)_3(PR_3)_2$ —	— $Fe(CO)_3(diars)$ $Fe(CO)(diars)_2$
$Cr(CO)_6$	$Cr(CO)_5(PR_3)$ $Cr(CO)_4(PR_3)_2$ $Cr(CO)_3(RCN)_3$ —	— $Cr(CO)_4(diars)$ — $Cr(CO)_2(diars)_2$

present in the dimers. The structure of $CpMn(CO)_3$ (BERNDT and MARSH[121]) is illustrated in formula B, and can be regarded as octahedral with the Cp group occupying three positions. (Cp = *cyclo*pentadiene, see below).

B

Similar structures are found for the chromium compounds $TCr(CO)_3$, where T = benzene (BAILEY and DAHL[122]), hexa-methylbenzene (BAILEY and DAHL[123]), phenanthrene (DEUSCHL and HOPPE[124]), and thiophene (BAILEY and DAHL[125]).

*cyclo*PENTADIENE COMPLEXES

The first example of this very interesting group of complexes to be discovered was 'ferrocene', an iron compound, $Fe(C_5H_5)_2$, where C_5H_5 is the *cyclo*pentadiene radical. A large number of similar compounds of general formula $M(C_5H_5)_2$, where M is one of the first row transition elements Ti—Ni, has now been made, and several of these compounds have been oxidized to the cation, $M(C_5H_5)_2{}^+$, in which M is formally tervalent. These compounds

Table 12.16.   cycloPentadiene Complexes Formed by the First Row Transition Metals

Element	Compound	m.p. °C	Colour	Magnetic moment (B.M.)	≡ Unpaired Electrons
Cu(II)	—	—	—	—	—
Ni(II)	Cp_2Ni	173	Green	2·86	2
Ni(III)	$[Cp_2Ni]^+$	—	Yellow	1·75	1
Co(II)	Cp_2Co	173	Purple	1·76	1
Co(III)	$[Cp_2Co]^+$	—	Yellow	Diamagnetic	0
Fe(II)	Cp_2Fe	173	Orange	Diamagnetic	0
Fe(III)	$[Cp_2Fe]^+$	—	Blue	2·26	1
Mn(II)	Cp_2Mn	173	Pink	5·9	5
Cr(II)	Cp_2Cr	173	Scarlet	2·84	2
Cr(III)	$[Cp_2Cr]^+$	—	Green	3·81	3
V(II)	Cp_2V	168	Purple	3·82	3
V(III)	$[Cp_2V]^+$	—	Purple	2·86	2
Ti(II)	Cp_2Ti	>130	Green	Diamagnetic	0
Ti(III)	$[Cp_2Ti]^+$	—	Green	2·3	1

are listed in *Table 12.16.* Several of the heavier transition metals, *e.g.* iridium and ruthenium, also form *cyclo*pentadiene derivatives similar to those in *Table 12.16*, and many other related compounds such as Cp_2TaBr_3 are known; readers interested in detailed accounts of these compounds are referred to recent review articles.[126, 127, 128]

*cyclo*Pentadiene compounds are soluble in the common organic solvents, and all the Cp_2M compounds (except Cp_2Ti) have melting points close to 173° C. The relatively simple infra-red spectra of all these compounds are very similar to each other, and indicate a high degree of symmetry. The visible and ultra-violet absorption spectra, on the other hand, are quite dissimilar, indicating very different electronic configurations. 'Ferrocene' was first thought to have a simple σ-bonded structure (C_5H_5)—Fe—(C_5H_5), but this formulation is inconsistent with the properties of the compound. An x-ray investigation shows that the molecule has the hitherto unknown 'sandwich' structure, where the iron atom is placed between the two *cyclo*pentadiene rings in a pentagonal-antiprismatical arrangement (*Figure 12.23*).

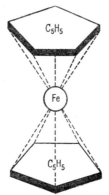

Figure 12.23. The 'sandwich' structure of ferrocene

While the arrangement of the atoms in ferrocene is known, there is a healthy measure of controversy about the nature of the bonding in the molecule, and quite a large number of theories has been advanced. It is not possible in a book of this size to discuss all these theories, some of which are extremely mathematical, and we shall limit ourselves to an elementary discussion in which we merely indicate some of the ideas behind the more quantitative theories.

We have to account for the bonding between the *cyclo*pentadiene radicals (C_5H_5) and the iron atom, and this bonding can be described as a pairing of an electron of each ring with one of the unpaired d electrons of the iron atom—so forming a single bond between the iron atom and each ring.

There are five atomic π-orbitals in each *cyclo*pentadiene ring, one on each carbon atom, and five molecular orbitals can be compounded from them. (We recall that in benzene, page 120, there are six atomic π-orbitals which give six molecular orbitals, three bonding and three anti-bonding.) The lowest energy molecular orbital (ψ_1) of *cyclo*pentadiene resembles the corresponding benzene orbital, with a continuous 'streamer' above and below the ring, extending over all five carbon atoms. The molecular orbitals of next lowest energy (ψ_2 and ψ_3) are degenerate, and resemble the 'split streamers' of benzene, with a node in either the xz or yz plane. (We take the line joining the centroids of the *cyclo*pentadiene rings and passing through the iron atom as the z axis; the x and y axes lie in a plane containing the iron atom parallel to the *cyclo*-pentadiene rings.) There are three electrons in the two orbitals ψ_2 and ψ_3, so that one of them must contain an unpaired electron, (*see Figure 12.24*).

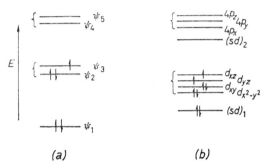

(a)   **(b)**

Figure 12.24. Relative energy levels for (*a*) π molecular orbitals of the C_5H_5 rings; (*b*) atomic orbitals (modified) of the iron atom. Degenerate levels, bracketed together, are separated for clarity

The electronic configuration of the ground state of iron is

Fe (argon core) $(3d)^2 (3d)^1 (3d)^1 (3d)^1 (3d)^1 (4s)^2$

MOFFITT[129] suggested that under the influence of the approaching *cyclo*pentadiene, radicals, the $4s$ and the $3d_{z^2}$ orbitals hybridize to give two non-equivalent sd hybrid orbitals (*see* earlier discussion on sd hybridization—page 238). One of these hybrid orbitals, ψ_2, denoted $(sd)_1$ has a lower energy than the $3d_{z^2}$ orbital, since the charge density is concentrated more in the xy plane, and repulsion with the electrons of the *cyclo*pentadiene is diminished; the other orbital, $(sd)_2$, is of higher energy, approximately that of the $4p$ orbitals, with an enhanced charge density around the z axis. Thus the modified configuration of the iron atom is written

Fe (argon core) $(sd)_1{}^2 (3d)^2 (3d)^1 (3d)^1$
$$(sd)_2{}^0 (4p_x)^0 (4p_y)^0 (4p_z)^0$$

We now get overlap between a molecular orbital of each ring with an appropriate d orbital of the iron atom (d_{xz} for one ring and d_{yz} for the other); one such overlap is illustrated in *Figure 12.25*. In some respects this bonding resembles that between olefins and metal ions (*see below*). However, this representation of bonding is oversimplified, and in the more sophisticated molecular-orbital

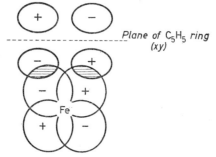

Figure 12.25. Overlap of Fe d_{xz} orbital with a π molecular orbital (shown in cross-section) of *cyclo*pentadiene

treatment the two bonds embrace the iron atom and both *cyclo*-pentadiene rings. Thus Moffitt obtained two delocalized bonding orbitals by combining the two iron d orbitals with a π molecular orbital (ψ_2 or ψ_3) from each ring; two anti-bonding orbitals are obtained at the same time:

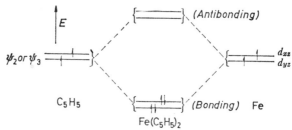

Other, still more elaborate treatments have been made (*see* WILKIN-SON and COTTON[126] for details) which allow for the participation of the $4p$ orbitals of iron and the π-bonding electrons of the rings in the bonding. All of the π electrons in the rings cannot take part completely in the bonding, however, since the aromatic character of the *cyclo*pentadiene rings is largely retained in the ferrocene-type compounds. Moreover magnetic anisotropy measurements on Cp_2Fe (MULAY and FOX[130]) indicate a single d_π—p_π bond between Fe and the rings.

We can use the Moffit scheme (*Figure 12.24*) to account for the magnetic properties of some of the compounds. Thus the extra electron in Cp_2Co (compared with Cp_2Fe) will go into either the $4p$ or the $(sd)_2$ orbitals, while in the corresponding nickel compound the two additional electrons will either singly occupy $4p$ orbitals or singly occupy a $4p$ and the $(sd)_2$ orbital. The chromium compound, with two electrons fewer than ferrocene, has the configuration $(sd)_1{}^2(3d)^1(3d)^1$, with two unpaired electrons. The scheme is not entirely satisfactory for vanadium and titanium, however, which have three unpaired and no unpaired electrons respectively, since to explain the three unpaired electrons in the vanadium compound it is necessary to assume that the $(sd)_1$ orbital has approximately the same energy as the $3d_{xy}$ and the $3d_{x^2-y^2}$ orbitals; this is not unreasonable, because the charge density in all three orbitals is largely concentrated in the xy plane, but it would predict two unpaired electrons for the titanium compound. More elaborate treatments proposed to account for this are outside the scope of this book.

We have not so far mentioned the manganese compound, but it is apparent that the magnetic moment, corresponding to five unpaired electrons, does not fit into the pattern of the Moffitt theory, which would predict one unpaired electron. The compound closely resembles the corresponding magnesium *cyclo*pentadiene, and evidently has an ionic structure containing $(C_5H_5)^-$ anions and Mn^{2+} cations (which have five unpaired electrons). The sandwich arrangement would be a natural one for three such ions.

Analogous compounds are formed by benzene, which contains one electron more than the *cyclo*pentadiene radical; thus $Cr(C_6H_6)_2$ is diamagnetic and has the configuration of ferrocene. There is some argument, as yet unresolved, about the detailed structure of *bis*benzenechromium, since JELLINEK[131] reports alternating bond lengths in the benzene rings, while COTTON and co-workers[132] find all the C—C lengths to be the same. IBERS[133] has re-examined COTTON's data and confirmed his findings, which are those to be expected on a simple bonding theory. Compounds with mixed ring systems such as $(C_5H_5)Cr(C_6H_6)$ have also been prepared.

OLEFIN COMPLEXES[134]

Ethylene, propylene and a number of other olefins form well-defined complexes with Pd(II), Pt(II), Cu(I) and Ag(I) salts. Divalent platinum, for instance, forms several compounds with ethylene: $K(C_2H_4PtCl_3)\cdot H_2O$, $(C_2H_4PtCl_2)_2$ and $(C_2H_4)_2PtCl_2$. The dimeric complex has the chlorine–chlorine bridged structure (I), while the compound with two ethylene groups per platinum atom has the *trans* configuration (II).

$$\text{I} \qquad\qquad \text{II}$$

A structure determination (ALDERMAN *et al.*[135]) of the related compound $Pt(C_2H_4 \cdot Cl_2 \cdot NHMe_2)$ has confirmed the generally accepted view that the two carbon atoms of the olefin are symmetrically linked to the platinum atom,

with the C—C bond perpendicular to the plane (say xy) of the remainder of the complex.

The bonding in complexes of this kind has long been discussed, and many suggestions have been advanced, but all earlier models involving non-equivalent bonding of the carbon atoms to platinum must be rejected in view of the results of x-ray work, and also because the results of the spectroscopic studies of Chatt and his co-workers show that the double bond of the olefin remains largely unchanged in the complex. The stretching frequency attributed to the C=C bond is merely lowered by some 143 cm^{-1} as compared with that of the olefin itself.

Ethylene evidently acts as a donor molecule, and 'feeds' two of its electrons into the vacant dsp^2 orbitals of the platinous ion to form a co-ordinate bond, although the exact origin of the donated electrons is not immediately clear. In any case, a simple donor–acceptor bond cannot be the answer because, if it were, ethylene would form similar but stronger complexes with acceptors such as the trihalides of boron; these complexes are very unstable, however, if they are formed at all.

It has been suggested (CHATT and DUNCANSON[136]), therefore, that platinum 'back-donates' electrons from its filled d orbitals into suitable vacant orbitals of the olefin molecule, so that there are two co-ordinate bonds between the olefin molecule and the platinum atom, arranged thus: $Pt \rightleftarrows C_2H_4$. This suggestion is in accord with the formation of other platinous complexes in which the donor molecules (PF_3 and AsR_3), have vacant d orbitals available for the back-donation of the electrons from the platinum atom.

A similar, but rather more elegant description by DEWAR[137] (for the Ag(I)-olefin complexes) has been extended by Chatt to include the platinous compounds, and it seems reasonable to apply it to all such complexes since they no doubt have a similar structure. A vacant dsp^2 hybrid orbital of the platinum atom is overlapped

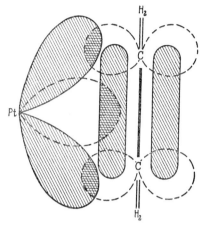

Figure 12.26. The orbital diagram for the double bonding of the olefin complexes of Pt(II) ions

by the π molecular orbital of ethylene to give a σ bond, and a π bond is formed by overlapping a doubly-filled dp hybrid orbital of the platinum atom with the vacant antibonding orbital of ethylene. A dp hybrid orbital is used rather than a simple d orbital since it gives a much better overlap, and, consequently, a more stable bond   The process is illustrated in *Figure 12.26*.

It is interesting to note that 1,5-hexadiene forms a compound, $C_6H_{10}PtCl_2$, which is much more stable than the analogous *trans-bis*ethylene compound (III). This stability would appear to be the result of the chelate nature of the compound, the *cis* configuration of which is in agreement with the dipole moment ($= 6D$).

III

Platinum also appears to form π-bonded acetylene derivatives such as $(Bu^tC{\equiv}CBu^t)_2Pt_2Cl_4$, in which the stretching frequency of the $C{\equiv}C$ bond is lowered by some 200 cm^{-1}. Other, more stable, acetylene derivatives of platinum have been prepared (Coates[138]) of the general type $(Ph_3P)_2Pt(CR \equiv CR)$, but these compounds appear not to be π-bonded. It seems more likely that the triple bond has opened so that each carbon atom may form a σ bond with the platinum atom. We can picture the Pt—C bonds being formed by the overlap of the dsp^2 orbitals of platinum with sp^2 orbitals of carbon; these bonds will be 'bent',

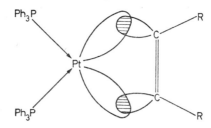

Figure 12.27. Acetylene complex of platinum(II)

since the sp^2 orbitals do not point directly towards the platinum atom, but there should, nevertheless, be appreciable overlap with the larger dsp^2 orbitals (*see Figure 12.27*).

THE REACTIVITY OF COMPLEX COMPOUNDS
(*cf.* Basolo and Pearson[4, 139]; Taube[140])

When we come to consider the reactivity of complex compounds we find that the rate at which complexes react is often more important than their stability; *i.e.* we consider the speed at which equilibrium is reached rather than the position of the equilibrium. Although a thermodynamically stable compound is often slow to react, and unstable complexes often react quickly, this is not always the case. Let us, for instance, consider reactions in which 'labelled' cyanide ions exchange with the unlabelled ions in some complexes. *Table 12.17* shows that the 4-co-ordinate complexes exchange cyanide ions rapidly, whereas the 6-co-ordinate complexes exchange only slowly. The exchange rates do not fall into the same order as the dissociation constants.

It should be noted that whereas the typical ionic reactions usually associated with inorganic compounds are instantaneous, complex compounds can react much more slowly. The reaction

Table 12.17.   Reactivity of Complex Cyanide Ions

Complex ion	Dissociation constant	Exchange rate with CN^-
$[Ni(CN)_4]^{2-}$	10^{-22}	Very fast
$[Fe(CN)_6]^{4-}$	10^{-37}	Very slow
$[Hg(CN)_4]^{2-}$	10^{-42}	Very fast
$[Fe(CN)_6]^{3-}$	10^{-44}	Very slow

rates vary widely, however, and depend upon the nature of the reactants and the experimental conditions, but complexes can be grouped into two fairly distinct 'reactivity classes':

1. Labile complexes, which react quickly (within the time of mixing).

2. Inert complexes, which react only slowly.

The distinction is surprisingly clear cut, and there are relatively few borderline cases. It is found that reactivity is very closely linked with the electronic configuration of the complex, and depends much less upon such other factors as the nature of the substituting ion, its charge, and its radius.

For the moment let us limit our discussion to octahedral complexes, and consider first the valence-bond and then the ligand-field approach to reactivity. The valence-bond method, we have seen, classifies the octahedral complexes into inner (d^2sp^3) and outer (sp^3d^2) complexes, and it is convenient to discuss these classes separately.

Inner Complexes (Strong-Field)

These complexes, which make use of the inner d orbitals, may be quite sharply divided into labile and inert compounds. If the metal ion has three or more electrons available for the d orbitals, it must, by Hund's rule, have at least one electron in each of the three d orbitals which are not used for bonding. Such ions are invariably inert. Those ions which have fewer than three electrons in the d level must have at least one completely vacant d orbital. Such ions are always labile.   *Table 12.18* lists examples of labile and inert complexes, together with the electronic configurations of the metal ions.

A very sharp change occurs at Cr^{3+}, whose complexes undergo only very slow substitution reactions in contrast to the rapid reactions of all V^{3+} complexes. The lability of complexes is quite evidently connected with the vacant d orbitals, and it seems likely

279

that in substitution reaction a vacant d orbital accepts electrons from the approaching group to form a 'transition state complex' with co-ordination number 7, which then breaks down releasing one of the originally co-ordinated groups, i.e.

$$\Upsilon + MX_6 = [\Upsilon MX_6] = \Upsilon MX_5 + X$$

Such a process (known in theoretical organic chemistry as an

Table 12.18.   Labile and Inert Inner-orbital Complexes

Ion	Nature of complexes	Electronic configuration*								
Sc³⁺, Ti⁴⁺	Labile	d^0	d^0	d^0	D^2	D^2	S^2	P^2	P^2	P^2
Ti³⁺	Labile	d^1	d^0	d^0	D^2	D^2	S^2	P^2	P^2	P^2
V³⁺	Labile	d^1	d^1	d^0	D^2	D^2	S^2	P^2	P^2	P^2
Cr³⁺	Inert	d^1	d^1	d^1	D^2	D^2	S^2	P^2	P^2	P^2
Mn³⁺	Inert	d^2	d^1	d^1	D^2	D^2	S^2	P^2	P^2	P^2
Fe³⁺	Inert	d^2	d^2	d^1	D^2	D^2	S^2	P^2	P^2	P^2
Co³⁺	Inert	d^2	d^2	d^2	D^2	D^2	S^2	P^2	P^2	P^2

* The electronic arrangement is described using Taube's nomenclature, the small d's representing orbitals not used in bonding and the capital D, S, and P's representing the orbitals that have accepted lone-pair electrons from the six ligands.

S_N2 process†), would readily take place, since the activation energy needed would be low.

With inert complexes, however, a vacant d orbital can only be made available by electron pairing or promotion, unless a high energy outer orbital is used for accepting electrons from the co-ordinating group, i.e.

$$Cr^{3+} \quad d^1 \ d^1 \ d^1 \ \boxed{D^2 \ D^2 \ S^2 \ P^2 \ P^2 \ P^2}$$

must change to either

$$d^2 \ d^1 \ d^0 \ \boxed{D^2 \ D^2 \ S^2 \ P^2 \ P^2 \ P^2}$$

or

$$d^1 \ d^1 \ d^0 \ \boxed{D^2 \ D^2 \ S^2 \ P^2 \ P^2 \ P^2} \ d^1 \ (\text{or } s^1)$$

† S_N2 denotes a bimolecular nucleophilic substitution reaction—a nucleophilic reagent being a donor molecule or group (e.g. NH_2^-)—while S_N1 denotes a unimolecular nucleophilic substitution reaction.

and

$$Co^{3+} \quad d^2 \ d^2 \ d^2 \ \boxed{D^2 \ D^2 \ S^2 \ P^2 \ P^2 \ P^2}$$

must change to

$$d^2 \ d^2 \ d^0 \ \boxed{D^2 \ D^2 \ S^2 \ P^2 \ P^2 \ P^2} \ d^1 \ d^1 \ (\text{or } s^2)$$

Any S_N2 process involving an intermediate complex of co-ordination number 7 would, in such cases, require a high activation energy, no matter what orbitals were used for the co-ordination of the substituting atom or group, and must, therefore, be very slow. It is found experimentally (BROWN and INGOLD[141]) that an S_N2 process only occurs with strongly nucleophilic reagents; with other reagents, the reaction proceeds by the S_N1 mechanism, *i.e.* one M—X bond breaks, leaving a transition state complex with five bonds, which then takes up the substituting group, Y. Thus

$$MX_6 \rightarrow [MX_5] + X$$

followed by

$$[MX_5] + Y \rightarrow YMX_5$$

The valence-bond approach gives only a qualitative picture of the relative reactivities of complexes, however, and it cannot compare two labile complexes with different configurations, *e.g.* d^1 and d^2. Basolo and Pearson have put the problem on a more quantitative basis by the application of ligand-field theory. They point out that when octahedral complexes react, the transition state in an S_N1 reaction will have 5-co-ordination (being either trigonal-bipyramidal or square-pyramidal), and in an S_N2 reaction will have 7-co-ordination (pentagonal-bipyramidal), and they assume that a change in C.F.S.E. in going from a co-ordination number of 6 to one of 5 or 7 can be considered as a ligand-field contribution to the activation energy of the reaction. If the C.F.S.E. for the transition state is smaller than that for the octahedral complex, the activation energy contribution will be large, and the reaction will be slow. *Table 12.19* summarizes these changes in C.F.S.E. for strong-field complexes; the differences quoted (in Δ) are those between the C.F.S.E. for the octahedral configuration and that for either the square-pyramidal or pentagonal-bipyramidal arrangements. Values are not given for transition states involving the trigonal-bipyramidal configuration since they are much greater than those for the square-pyramidal arrangement; an S_N1 reaction would therefore always go by way of the square-pyramidal transition state.

Table 12.19.   Losses in C.F.S.E. for Strong-Field Octahedral Complexes (values quoted in Δ)

Number of d electrons	Square-pyramidal transition state (S_N1)	Pentagonal-bipyramidal transition state (S_N2)
0	0	0
1	$-0 \cdot 06$	$-0 \cdot 13$
2	$-0 \cdot 11$	$-0 \cdot 26$
3	$0 \cdot 20$	$0 \cdot 43$
4	$0 \cdot 14$	$0 \cdot 30$
5	$0 \cdot 09$	$0 \cdot 17$
6	$0 \cdot 40$	$0 \cdot 85$
7	$-0 \cdot 11$	$0 \cdot 53$
8	$0 \cdot 20$	$0 \cdot 43$
9	$-0 \cdot 31$	$0 \cdot 11$
10	0	0

From the values quoted in *Table 12.19,* it can be seen that irrespective of whether the reaction proceeds by an S_N1 or an S_N2 mechanism, the d^0, d^1, and d^2 complexes will react quickly compared with the d^3, d^4, d^5, and d^6 complexes.   Thus the ligand-field approach confirms the general predictions of the valence-bond method, but the C.F.S.E. changes quoted in *Table 12.19* further suggest a decreasing order of reactivity:

$$d^5 > d^4 > d^3 > d^6$$

Outer Complexes (Weak-Field)

These also may be labile or inert, but, unlike the inner-orbital complexes, there is no sharp dividing line. In general, labile complexes are formed by metal ions which use outer orbitals, *e.g.* Al^{3+}, Mn^{2+}, Fe^{2+}, Co^{2+}, Zn^{2+}, Cd^{2+}, Hg^{2+}, Ga^{3+}, In^{3+}, Tl^{3+}. The halogen complexes of the non-metals and metalloids are usually fairly inert; the following may be cited: $[SiF_6]^{2-}$, $[PF_6]^-$, $[AsF_6]^-$, $[SbCl_6]^-$, SF_6.

For any given isoelectronic series, the lability decreases with an increase in the valency of the central atom

e.g. $[AlF_6]^{3-}$ $[SiF_6]^{2-}$ $[PF_6]^-$ SF_6

| Very rapid hydrolysis | Fairly rapid hydrolysis | Slow hydrolysis | Very slow hydrolysis |

In such a series, the rate of hydrolysis becomes less as the valency of the central atom increases, and it seems likely that the substitution proceeds by an S_N1 mechanism, the complex first losing a ligand and then taking up the substituting one. This is probable, since the removal of one or two ligands from a complex such as $[AlF_6]^{3-}$ should not involve a large energy change, because the aluminium atom no longer makes use of the high-energy d orbitals in bonding. With a central atom of larger charge, however, the bonds are more covalent and they are less easily broken. A high activation energy will be expected, therefore, whether the first step of the mechanism is one of bond breaking (S_N1) or of co-ordination of the substituting group (S_N2).

Although in this series the lability decreases as the bonds become more covalent, it must be emphasized that a highly covalent bond does not necessarily imply an inert complex in other cases. Molybdenum(V) and tungsten(V) complexes, for instance, are more labile than complexes of the tervalent elements, even though the bonds are more covalent.

We can again use the ligand-field approach to get a measure of the C.F.S.E. change involved in the formation of either a 5- or 7-co-ordinate transition state; these results are summarized in *Table 12.20*. These values show that all the configurations should lead to fast reactions except d^3 and d^8. The ligand-field approach explains the reactivity by saying that there is no loss of C.F.S.E. in forming the transition state, whereas the valence-bond approach accounts for the lability by saying that since high-energy orbitals are used in the bonding, the bonds will be more easily broken (giving an S_N1 mechanism).

Table 12.20. Losses in C.F.S.E. for Weak-Field Octahedral Complexes (values quoted in Δ)

Number of d electrons	Square-pyramidal transition state (S_N1)	Pentagonal-bipyramidal transition state (S_N2)
0, 5, 10	0	0
1, 6	$-0 \cdot 06$	$-0 \cdot 13$
2, 7	$-0 \cdot 11$	$-0 \cdot 26$
3, 8	$0 \cdot 20$	$0 \cdot 43$
4, 9	$-0 \cdot 31$	$0 \cdot 11$

The two approaches disagree over the complexes with a d^8 configuration (*e.g.* Ni(II)). The valence-bond method suggests a lability similar to that for other complexes with any configuration from d^5 to d^9, but the ligand-field calculations indicate that the d^8 complexes should be inert and comparable to d^3 (strong-field) complexes. There is quite a bit of evidence to show that, in comparison with complexes of other divalent ions, those of nickel react more slowly. It must be emphasized, however, that Ni(II) complexes cannot be compared with those of Cr(III), (which have the d^3 configuration), because Δ is much larger for a tervalent ion.

THE REACTIVITY OF COMPLEXES OF CO-ORDINATION NUMBER 4 (THE 'TRANS-EFFECT')[142]

It has been shown that complexes making use of sp^3 orbitals for bonding, such as $[HgX_4]^{2-}$ ($X = $ Cl, Br, and I) are labile, which might be expected since the outer d orbitals are energetically available for accepting electrons from the substituting ligand. Moreover, from a ligand-field point of view, there is no C.F.S.E. in the d^{10} configuration of Hg(II), and therefore no loss in forming a transition state of co-ordination number 5. Some square-planar complexes are labile (e.g. $[Ni(CN)_4]^{2-}$), and others are inert (e.g. $[PtX_4]^{2-}$), although valence-bond considerations would predict them all to be labile because of the available p orbital. However, the distinctions become easier to understand if ligand-field splittings are considered. Thus, for a given type of ligand, the value of Δ is very much greater for platinum than for nickel, so that the loss in C.F.S.E. in forming the transition state will be accordingly greater.

A great deal of work has been done on the substitution reactions of square-planar complexes, especially by the Russian school and by Chatt and co-workers. They have shown that a negative ligand (e.g. Cl⁻) has a noticeable effect on the reactivity of the bond

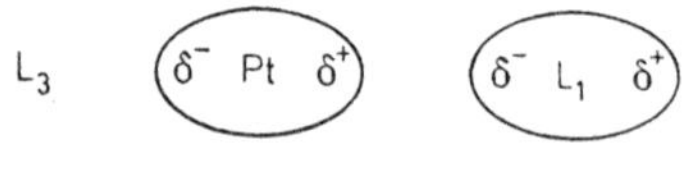

$$L_3 \quad \left(\delta^- \; Pt \; \delta^+ \right) \quad \left(\delta^- \; L_1 \; \delta^+ \right)$$

Figure 12.28. Mutual polarization between Pt and L_1

'*trans*' to it, making it much more labile, but has no such effect on '*cis*' bonds. This labilizing effect is referred to as the '*trans*-effect'. From various qualitative observations, the ligands can be placed in an order of increasing *trans*-effect:

$$H_2O < OH^- < NH_3 < py < Cl^- < Br^- < I^- < CN^-$$

Two general explanations have been put forward to account for the *trans*-effect; the first one is essentially electrostatic and suggests a weakening of the *trans* bond; the second one proposes π-bonding between the ligand and platinum, which influences the *trans* position but not the *cis* position.

In the simple electrostatic approach, polarization effects are considered. Thus the mutual polarization of ligand L_1 and the platinum atom in $PtL_1L_2L_3L_4$ can be represented by *Figure 12.28*. If the four ligands are the same, then these mutual polarization effects cancel out, but in a case where, for instance, L_1 differs from the other three ligands, then L_2 and L_4 balance each other, but L_1 and L_3 do not. If L_1 is more polarizable than the other ligands, then the overall effect is to put a partial negative charge on the side of Pt nearest to L_3. This *trans*-effect weakens the Pt—L_3 bond. Hence for a series of ligands we should expect to get the largest *trans*-effect with the ligand that is the most easily polarized.

This order is in fact found, but only for ligands which cannot take part in π bonding, *e.g.* H_2O and NH_3. In a series of Pt(II) complexes of the type $PtLACl_2$, where $A = NH_3$ or amines, and L is a variety of ligands (*trans* to A) that cannot form π bonds, the infra-red spectra show an inverse correlation between the Pt—N bond strength (inferred from the stretching frequency) and the electrostatic *trans*-effect of L.

Electrostatic calculations with π-bonding ligands lead to the wrong order. With such ligands, *e.g.* C_2H_4, and PR_3, which have a very high *trans*-effect, we assume that the doubly-filled d orbitals of the platinum atom give d_π—d_π bonding (*see* page 215), and that

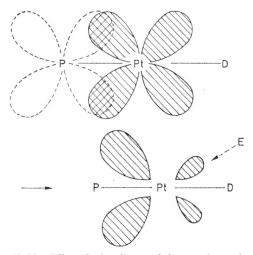

Figure 12.29. Effect of π bonding on d-electron charge density

the effect of such π bonding is to alter the d-electron charge density by dragging it into the region between the ligand (P) and the platinum atom. The gap which is created in the d-electron charge cloud permits a more ready approach of the substituting ligand (E) (*see Figure 12.29*). If the four ligands are in the xy plane, then the phosphorus atom can π bond with the doubly-filled d_{xz} orbital, and the approaching ligand (E) approaches in the xz plane giving a transition state with P, D, and E all in this plane (*cf. Figure 12.30*).

The π bonding also strengthens the transition-state complex, thus lowering the activation energy and speeding up the reaction.

This π-bonding description accounts admirably for the *trans-*

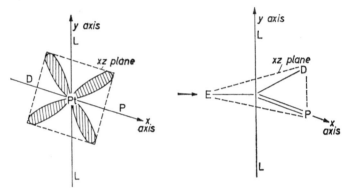

Figure 12.30. Formation of transition state in substitution reactions of *square-planar* complexes

effect order observed for the π-bonding ligands. With ligands such as Cl^-, Br^-, I^-, and CN^-, both π-bonding and electrostatic factors must be taken into consideration; a suitable weighting of these two factors can account for the observed *trans*-effect.

REFERENCES

[1] BAILAR, J. C. *The Chemistry of the Co-ordination Compounds*, Reinhold, New York, 1956

[2] COLLINGE, R. N., NYHOLM, R. S. and TOBE, M. L. *Nature, Lond.* 201 (1964) 1322

[3] KIMBALL, G. E. *J. chem. Phys.* 8 (1940) 188

[4] BASOLO, F. and PEARSON, R. G. *Mechanisms of Inorganic Reactions*, p. 41, Wiley, New York, 1958

[5] GRAY, H. B. *J. chem. Educ.* 41 (1964) 2

[6] ORGEL, L. E. *An Introduction to Transition Metal Chemistry: Ligand-field Theory*, p. 41, Methuen, London, 1960

[7] COTTON, F. A. *J. chem. Educ.* 41 (1964) 466

REFERENCES

[8] BALLHAUSEN, C. J. *Introduction to Ligand-field Theory*, McGraw-Hill, New York, 1962

[9] JØRGENSEN, C. K. *Absorption Spectra and Chemical Bonding in Complexes.* Pergamon, Oxford, 1962

[10] RUSS, B. J. and FOWLES, G. W. A. *Chem. Comm* (1966) 19

[11] FOWLES, G. W. A., HOODLESS, R. A. and WALTON, R. A. *J. chem. Soc.* (1963) 5873

[12] FOWLES, G. W. A. and WALTON, R. A. *J. chem. Soc.* (1964) 4953

[13] FOWLES, G. W. A. and HOODLESS, R. A. *J. chem. Soc.* (1963) 33

[14] FIGGIS, B. N. and LEWIS, J. *Prog. inorg. Chem.* 6 (1964) 37

[15] BURSTALL, F. H. and NYHOLM, R. S. *J. chem. Soc.* (1952) 3570

[16] CRAIG, D. P., MACCOLL, A., NYHOLM, R. S., ORGEL, L. E. and SUTTON, L. E. *J. chem. Soc.* (1954) 322

[17] WILLIAMS, R. J. P. *Annual Reports* 56 (1959) 87

[18] BJERRUM, J. *Stability Constants*, The Chemical Society, London, 1957

[19] GRIFFITH, J. S. and ORGEL, L. E. *Q. Rev. chem. Soc.* 11 (1957) 381

[20] GUZZETTA, F. H. and HADLEY, W. B. *Inorg. Chem.* 3 (1964) 259

[21] WOOD, J. L. and JONES, M. M. *Inorg. Chem.* 3 (1964) 1553

[22] GEORGE, P. and McCLURE, D. S. *Prog. inorg. Chem.* 1 (1959) 381

[23] BERG, R. A. and SINANOGLU, O. *J. chem. Phys.* 32 (1960) 1082

[24] KNOX, K. *Acta crystallogr.* 14 (1961) 583

[25] MOROSIN, B. and GRAEBER, E. J. *J. chem. Phys.* 42 (1965) 898

[26] MOROSIN, B. and GRAEBER, E. J. *Acta crystallogr.* 16 (1963) 1176

[27] ORGEL, L. E. *An Introduction to Transition Metal Chemistry. Ligand-field Theory*, p. 60. Methuen, London, 1960

[28] TRACY, J. W., GREGORY, N. W., LINGAFELTER, E. C., DUNITZ, J. D., MEZ, H. C., RUNDLE, R. E., SCHERINGER, C., YAKEL, H. L. and WILKINSON, M. K. *Acta crystallogr.* 14 (1961) 927

[29] HEPWORTH, M. A. and JACK, K. H. *Acta crystallogr.* 10 (1957) 345

[30] STEINFINK, H. and BURNS, J. H. *Acta crystallogr.* 17 (1964) 823

[31] MOROSIN, B. and BRATHOVDE, J. R. *Acta crystallogr.* 17 (1964) 705

[32] STEPHENSON, N. C. *J. inorg. nucl. Chem.* 24 (1962) 791, 797; *Acta crystallogr.* 17 (1964) 592

[33] FISHER, P. J., TAYLOR, N. E. and HARDING, M. M. *J. chem. Soc.* (1960) 2303

[34] BROWN, I. D. and DUNITZ, J. D. *Acta crystallogr.* 14 (1961) 480

[35] ORGEL, L. E. *An Introduction to Transition-Metal Chemistry: Ligand-field Theory*, p. 66. Methuen, London, 1960

[36] VOSSOS, P. H., FITZWATER, D. R. and RUNDLE, R. E. *Acta crystallogr.* 16 (1963) 1037

[37] WILLETT, R. D., WIGGINS, C. D., KRUSH, R. F. and RUNDLE, R. E. *J. chem. Phys.* 38 (1963) 2429

[38] SCHNERING, H. G. *Naturwissenschaften* 48 (1961) 665

[39] LANE, A. P. and PAYNE, D. S. *J. chem. Soc.* (1963) 4004

[40] CASS, R. C., COATES, G. E. and HAYTER, R. G. *J. chem. Soc.* (1955) 4007

[41] AHRLAND, S. and CHATT, J. *Chemy Ind.* (1955) 96

[42] COTTON, F. A. *J. chem. Soc.* (1960) 5268

[43] DAHL, L. F. and WAMPLER, D. L. *Acta crystallogr.* 15 (1962) 903

[44] HOLM, R. H. and COTTON, F. A. *J. chem. Phys.* 32 (1960) 1168

[45] FORRESTER, J. D., ZALKIN, A. and TEMPLETON, D. H. *Inorg. Chem.* 3 (1964) 1500

[46] DAVISON, A., EDELSTEIN, N., HOLM, R. H. and MAKI, A. H. *J. Am. chem. Soc.* 85 (1963) 3049

[47] COTTON, F. A. and ELDER, R. C. *J. Am. chem. Soc.* 86 (1964) 2294

[48] COTTON, F. A. and SOLDEBERG, R. H. *J. Am. chem. Soc.* 85 (1963) 2402

[49] ELDER, M. and PENFOLD, B. R. *Chem. Comm.* (1965) 308

[50] ROBINSON, W. T., FERGUSON, J. E. and PENFOLD, B. R. *Proc. chem Soc.* (1963) 116

[51] BERTRAND, J. A., COTTON, F. A. and DOLLASE, W. A. *J. Am. chem. Soc.* 85 (1963) 1349

[52] COTTON, F. A. and MAGUE, J. T. *Inorg. Chem.* 3 (1964) 1094

[53] STEPHENSON, N. C. *Acta crystallogr.* 17 (1964) 587

[54] FRASSON, E., PANATTONI, C. and SACCONI, L. *Acta crystallogr.* 17 (1964) 85, 477

[55] BULLEN, G. J., MASON, R. and PAULING, P. *Inorg. chem.* 4 (1965) 456

[56] FRASSON, E. and PANATTONI, C. *Acta crystallogr.* 13 (1960) 893

[57] EISENBERG, R. and IBERS, J. A. *Inorg. Chem.* 4 (1965) 605

[58] GARTON, G., HENN, D. E., POWELL, H. M. and VENANZI, L. M. *J. chem. Soc.* (1963) 3625

[59] GILL, N. G. and NYHOLM, R. S. *J. chem. Soc.* (1959) 3997

[60] GRUEN, D. M. *Q. Rev. chem. Soc.* 19 (1965) 354

[61] FRASSON, E., PANATTONI, C. and SACCONI, L. *J. phys. Chem. Ithaca* 63 (1959) 1908

[62] FOX, M. R., ORIOLI, P. L., LINGAFELTER, E. C. and SACCONI, L. *Acta crystallogr.* 17 (1964) 1159

[63] CHAKRAVORTY, A., FENNESSY, J. P. and HOLM, R. H. *Inorg. Chem.* 4 (1965) 26

[64] LA VILLA, R. E. and BAUER, S. H. *J. Am. chem. Soc.* 85 (1963) 3597

[65] JARSKI, M. A. and LINGAFELTER, E. C. *Acta crystallogr.* 17 (1964) 1109

[66] ORIOLI, P. L., LINGAFELTER, E. C. and BROWN, B. W. *Acta crystallogr.* 17 (1964) 1113

[67] MOROSIN, B. and LINGAFELTER, E. C. *Acta crystallogr.* 13 (1960) 807

[68] FERGUSON, J. *J. chem. Phys.* 40 (1964) 3406

[69] WILLETT, R. D. *J. chem. Phys.* 41 (1964) 2243

[70] GILLESPIE, R. J. *J. chem. Soc.* (1963) 4672, 4679

[71] FOWLES, G. W. A. and HOODLESS, R. A. *J. chem. Soc.* (1963) 33

[72] BEATTIE, I. R. and GILSON, T. *J. chem. Soc.* (1965) 6595

[73] DODGE, R. P., TEMPLETON, D. H. and ZALKIN, A. *J. chem. Phys.* 35 (1961) 55

[74] HAZELL, A. C. *J. chem. Soc.* (1963) 5745

[75] DUCKWORTH, M. W., FOWLES, G. W. A. and WILLIAMS, R. G. *Chemy Ind.* (1962) 1285

[76] COTTON, F. A. and HARRIS, C. B. *Inorg. Chem.* 4 (1965) 330

[77] LA PLACA, S. J. and IBERS, J. A. *Inorg. Chem.* 4 (1965) 778

[78] MAIR, G. A., POWELL, H. M. and HENN, D. E. *Proc. chem. Soc.* (1960) 415

[79] COTTON, F. A., DUNNE, T. G. and WOOD, J. S. *Inorg. Chem.* 4 (1965) 318

[80] DYER, G., HARTLEY, J. G. and VENANZI, L. M. *J. chem. Soc.* (1965) 1293

[81] MAIR, G. A., POWELL, H. M. and VENANZI, L. M. *Proc. chem. Soc.* (1961) 170

[82] MORI, M., SAITO, Y. and WATANABE, T. *Bull. chem. Soc.*, Japan 34 (1961) 295

[83] CORBRIDGE, D. E. C. and COX, E. G. *J. chem. Soc.* (1956) 594

[84] MONTGOMERY, H. and LINGAFELTER, E. C. *Acta crystallogr.* 16 (1963) 748

[85] BARCLAY, G. A. and KENNARD, C. H. L. *Nature, Lond.* 192 (1961) 425

[86] BALLHAUSEN, C. J. and GRAY, H. B. *Inorg. Chem.* 1 (1962) 111

[87] BRITTON, D. *Can. J. Chem.* 41 (1963) 1632

[88] GANORKAR, M. C. and STIDDARD, M. H. B. *J. chem. Soc.* (1965) 3494

[89] NIGAM, H. L., NYHOLM, R. S. and STIDDARD, M. H. B. *J. chem. Soc.* (1960) 1808

[90] CLARK, R. J. H., KEPERT, D. L. and NYHOLM, R. S. *J. chem. Soc.* (1965) 2877

[91] CLAASEN, H. H. and SELIG, H. *J. chem. Phys.* 43 (1965) 103

[92] CLARK, R. J. H., LEWIS, J. and NYHOLM, R. S. *J. chem. Soc.* (1962) 2460

[93] ADDISON, C. C., GARNER, C. D., SIMPSON, W. B., SUTTON, D. and WALLWORK, S. C. *Proc. chem. Soc.* (1964) 367

[94] CLARK, R. J. H., KEPERT, D. L., LEWIS, J. and NYHOLM, R. S. *J. chem. Soc.* (1965) 2865

[95] HOARD, J. L. and SILVERTON, J. V. *Inorg. Chem.* 2 (1963) 235

[96] RANDIĆ, M. *J. chem. Phys.* 36 (1962) 2094

[97] KEPERT, D. L. *J. chem. Soc.* (1965) 4736

[98] COTTON, F. A. and BERGMAN, J. G. *J. Am. chem. Soc.* 86 (1964) 2941

[99] GRDENIĆ, D. and MATKOVIC, B. *Acta crystallogr.* 12 (1959) 817

[100] WOLF, L. *Acta crystallogr.* 13 (1960) 778

[101] ZALKIN, A., FORRESTER, J. D. and TEMPLETON, D. H. *Inorg. Chem.* 3 (1964) 639

[102] FLEMING, J. E. and LYNTON, H. *Chemy Ind.* (1959) 1409; (1960) 1416

[103] ABEL, E. W. *Q. Rev. chem. Soc.* 17 (1963) 133

[104] DONOHUE, J. and CARON, A. *Acta crystallogr.* 17 (1964) 663

[105] DAVIS, M. I. and HANSON, H. P. *J. phys. Chem.* 69 (1965) 3405

[106] CALDERAZZO, F., CINI, R., CORRADINI, P., ERCOLI, R., and NATTA, G. *Chemy Ind.* (1960) 500

[107] DAHL, L. F., ISHISHI, E. and RUNDLE, R. E. *J. chem. Phys.* 26 (1957) 1750; BAILEY, M. F. and DAHL, L. F. *Inorg. Chem.* 4 (1965) 1140

[108] SUMNER, G. G., KING, H. P. and ALEXANDER, L. E. *Acta crystallogr.* 17 (1964) 732

[109] COREY, E. R. and DAHL, L. F. *Inorg. Chem.* 1 (1962) 521

[110] DAHL, L. F. and RUNDLE, R. E. *J. chem. Phys.* 26 (1957) 1751

[111] DOBSON, S. R. and SHELINE, R. K. *Inorg. Chem.* 2 (1963) 1313

[112] HERBER, R. H., KINGSTON, W. R. and WERTHEIM, G. K. *Inorg. Chem.* 2 (1963) 153

[113] DAHL, L. F. and BLOUNT, J. F. *Inorg. Chem.* 4 (1965) 1373

[114] SMITH, D. L. *J. chem. Phys.* 42 (1965) 1460

[115] COREY, E. R., DAHL, L. F. and BECK, W. *J. Am. chem. Soc.* 85 (1963) 1202

[116] EDGELL, W. F. and GALLUP, G. *J. Am. chem. Soc.* 78 (1956) 4188

[117] EDGELL, W. F., ASATO, G., WILSON, W. and ANGELL, G. *J. Am. chem. Soc.* 81 (1959) 2022

[118] Based on a number of papers, particularly of Nyholm and co-workers. Original references may be found in *Annual Reports* (1950–64)

[119] KILBOURN, B. T., BLUNDELL, T. L. and POWELL, H. M. *Chem. Comm.* (1965) 444

[120] PAUSON, P. L. *Tilden Lecture. Proc. chem. Soc.* (1960) 297

[121] BERNDT, A. F. and MARSH, R. E. *Acta crystallogr.* 16 (1963) 118

[122] BAILEY, M. F. and DAHL, L. F. *Inorg. Chem.* 4 (1965) 1314

[123] BAILEY, M. F. and DAHL, L. F. *Inorg. Chem.* 4 (1965) 1298

[124] DEUSCHL, H. and HOPPE, W. *Acta crystallogr.* 17 (1964) 800

[125] BAILEY, M. F. and DAHL, L. F. *Inorg. Chem.* 4 (1965) 1306

[126] WILKINSON, G. and COTTON, F. A. *Prog. inorg. Chem.* 1 (1959) 1

[127] FISCHER, E. O. and FRITZ, H. P. *Adv. inorg. Chem. Radiochem.* 1 (1959) 56

[128] RAUSCH, M. D. *Can. J. Chem.* 41 (1963) 1289

[129] MOFFITT, W. *J. Am. chem. Soc.* 76 (1954) 3386

[130] MULAY, L. N. and FOX, M. E. *J. chem. Phys.* 38 (1963) 760

[131] JELLINEK, F. *J. organomet. Chem.* 1 (1963) 43

[132] COTTON, F. A., DOLLASE, W. A. and WOOD, J. S. *J. Am. chem. Soc.* 85 (1963) 1543

[133] IBERS, J. A. *J. chem. Phys.* 40 (1964) 3129

[134] GUY, R. G. and SHAW, B. L. *Adv. inorg. Radiochemistry* 4 (1962) 78

[135] ALDERMAN, P. R. H., OWSTON, P. G. and ROWE, J. M. *Acta crystallogr.* 13 (1960) 149

[136] CHATT, J. and DUNCANSON, L. A. *J. chem. Soc.* (1953) 2939

[137] DEWAR, M. J. S. *Bull. Soc. Chim.* 18 (1951) C79

[138] COATES, G. E. *Organometallic Compounds.* Second Edition. p. 347. Methuen, London, 1960

[139] BASOLO, F. and PEARSON, R. G. *Adv. inorg. Chem. Radiochem.* 3 (1961) 1

[140] TAUBE, H. *Chem. Rev.* 50 (1952) 69

[141] BROWN, D. D. and INGOLD, C. K. *J. chem. Soc.* (1953) 2680

[142] BASOLO, F. and PEARSON, R. G. *Prog. inorg. Chem.* 4 (1962) 381

13

ELECTRON-DEFICIENT MOLECULES

INTRODUCTION

WE have seen that some elements of groups II and III—namely beryllium, boron, and aluminium—have a considerable tendency to accept electrons, and so acquire a tetrahedral configuration whenever possible. Typical examples discussed in Chapter 11, such as $BMe_3 \cdot NH_3$, are substances in which the acceptor atom receives electrons from a donor with a suitable 'lone pair', thus producing a compound with normal two-electron bonds, one being represented formally as a co-ordinate bond.

We have also seen that this tendency to achieve a co-ordination number of four is so great that in the absence of suitable donor molecules, simple compounds of these elements may dimerize, e.g. $2 \, AlCl_3 \rightarrow Al_2Cl_6$. Normal two-electron bonds are present in such halogen-bridged dimers. It is more difficult to write down structural formulae for dimers such as diborane, B_2H_6, and the alkyls of aluminium and beryllium, $Al_2(CH_3)_6$ and $Be_2(CH_3)_4$, since there are not enough electrons to provide two-electron bonds between the atoms in these molecules. Such molecules are, therefore, said to be 'electron-deficient'. We shall first discuss the structure of diborane in some detail.

THE STRUCTURE OF DIBORANE (*cf.* LONGUET-HIGGINS[1])

This structural problem, which has been a source of controversy for many years, can be tackled in two stages—firstly, we must consider the relative positions of the atomic nuclei concerned, and, secondly, we must give a theoretical interpretation of the bonding in any particular nuclear configuration.

The Nuclear Configuration of Diborane

The two main configurations which have been put forward are the so-called 'ethane' and 'bridge' models

'ethane'   'bridge'

291

The 'ethane' structure is analogous to that of C_2H_6, and the 'bridge' structure is comparable with that of Al_2Cl_6. The ethane model formed the basis of earlier discussions of the diborane structure, but more recently it has been superseded by the bridge model. In either case there are not enough electrons to provide electron-pair links, *i.e.* both structures are electron deficient.

We can also write the bridge structure as

$$\begin{array}{c}
\text{H}^+ \\
\text{H} \diagdown \quad \quad \diagup \text{H} \\
\quad \bar{\text{B}} = = = = = \bar{\text{B}} \\
\text{H} \diagup \quad \quad \diagdown \text{H} \\
\text{H}^+
\end{array}$$

i.e. as an ethylene-like molecule with a double bond between formally negative B atoms. The position of the two protons was unspecified in earlier models, but recent speculations have placed them between the two B atoms, so that the nuclear configuration is more or less identical with that of the bridge model proper. This ethylene-like structure will assume a greater significance when the bonding is discussed, but for the moment it will be considered simply as a bridge structure.

The essential difference between the two models is that whereas all six hydrogen atoms are equivalent in the ethane structure, two have a special position in the bridge model. Moreover, free rotation about the boron–boron axis should be possible for the ethane model but not for the bridge structure. It is significant that only four of the six hydrogen atoms in B_2H_6 can be replaced by methyl groups, giving $Me_4B_2H_2$, and that this tetramethyl molecule is symmetrical with two methyl groups attached to each boron atom. This suggests that two of the six hydrogen atoms in diborane are attached differently from the other four.

Extensive physical measurements have been made on diborane:

(a) Early x-ray diffraction work was of little help because only the boron atoms could be located, but SMITH and LIPSCOMB[2] have recently studied βB_2H_6 and found it to have a bridge structure with terminal and bridging B—H distances of 1·10 and 1·25 Å respectively.

(b) electron-diffraction measurements are more useful, but inasmuch as the two models differ mainly in their H—H distances, and since the scattering produced by H atoms is small, the results

292

are difficult to interpret. A recent investigation by Hedberg and Schomaker has decided unambiguously in favour of the bridge structure with the dimensions shown in the diagram.

A further study by BARTELL and CARROLL[3] has confirmed this structure for both diborane and deuterodiborane, with terminal and bridging B—H linkages of 1·20 and 1·34 Å respectively. Two boron atoms and the four terminal hydrogen atoms are co-planar, and the two bridge hydrogens lie above and below the plane. A significant point is the difference between the bridge and terminal B—H bond lengths. Thus whereas the terminal B—H distances are the same as that found for a normal single B—H bond, as in $BH_3 \cdot CO$ (1·20 Å), the bridge B—H bonds are much longer and hence weaker.

An electron-diffraction study of tetramethyldiborane shows the carbon atoms of the four methyl groups to be co-planar with the two boron atoms, agreeing with a bridge formulation.

(c) the Raman and infra-red vibration spectra of diborane agree well with the spectra expected for an ethylene-like structure (i.e. a bridge structure). Two intense frequencies (2,102 and 2,523 cm^{-1}) in the Raman spectrum can be assigned to B—H bond-stretching vibrations. Now the ethane model provides for only one such symmetrical stretching vibration, while the bridge model suggests two—the stretching of the B—H bonds in the bridge, and the stretching of the terminal B—H bonds. Eight funda-mental bands are found in the infra-red spectrum, only five of which can be accounted for on the basis of an ethane model, but all would be expected in a bridge-type molecule. One band in particular, that at 412 cm^{-1}, cannot be explained satisfactorily by an ethane model, but may be readily interpreted as a bending of the bridge molecule about a line joining the two bridge hydrogen atoms.

(d) the measured specific heat of diborane is different from that calculated for an ethane model (STITT[4]); this discrepancy can only be explained by assuming restricted rotation of the two BH_3 groups. This restriction requires an energy barrier of from 5 to 15 kcal compared with 3 kcal for ethane itself, which is rather unlikely in view of the longer B—B distance (B—B = 1·77 Å, C—C = 1·54 Å). Very much better agreement results if a bridge model is assumed.

(e) particularly convincing evidence for the bridge structure is provided by nuclear magnetic resonance studies (SHOOLERY[5]) on isotopically labelled diborane, $^{11}B_2H_6$. Three main regions of absorption are found, one associated with the boron atoms, one with the terminal hydrogen atoms, and one with the bridge hydrogen atoms. The fine structure of these absorption bands, which results from interactions between neighbouring nuclei, is just what is predicted for the bridge model.

The physical evidence thus shows quite definitely that the nuclear configuration is bridge-like, and we must now try to explain the nature of the bonding in such a molecule. The interpretation of bonding is, of course, much more a matter of conjecture, and it seems unlikely that the last word has yet been said; nevertheless a quite reasonable electronic structure can be put forward.

The Bonding in Diborane

The terminal B—H links are of the length expected for normal single B—H bonds and are, accordingly, assumed to be of this type. In these circumstances the electron deficiency must be in the bridge bonds, and we shall review several suggested interpretations of these curious linkages.

(a) BURAWOY[6] has suggested that hydrogen bonding might be responsible for holding together the two BH_3 groups. Thus although the tendency for hydrogen bonding decreases in the series $F > O > N > C$, Burawoy postulates that it may increase again with boron, the hydrogen now being at the negative end of the dipole. This seems very unlikely, because the B—H bond is virtually without a dipole moment (the electronegativity values are $B = 2 \cdot 0$, $H = 2 \cdot 1$). Even if we accept this hypothesis of a B ... H ... B hydrogen bond, there are still other serious objections to the assumption that these links are responsible for the bonding in diborane. We have seen in Chapter 10 that hydrogen bonds are very weak, and also, that the hydrogen atom is not, in general, equidistant from the two atoms linked by it. In diborane, however, the bond is fairly strong, and the electron-diffraction and molecular-spectrum investigations show that the bridge H atoms are equidistant from the two B atoms. We can therefore eliminate electrostatic bonding of this type.

(b) Diborane has been described as a resonance hybrid of the covalent forms I and II, thus producing a symmetrical structure.

I II

SYRKIN and DYATKINA[7] have calculated that the resulting resonance energy would be too small to compensate both for the unfavourable bond angles and for the repulsion of electrons not taking part in bonding. Further calculations by Syrkin and Dyatkina, incorporating such ionic forms as III and IV, suggest that sufficient resonance energy might then be available to stabilize the molecule.

III   IV

However, the bond angles in these ionic forms would be considerably distorted, $\overset{+}{B}H_2$ being normally linear (sp), BH_3 planar (sp^2) and BH_4^- tetrahedral (sp^3). These unfavourable bond angles, as well as the repulsion between non-paired electrons (which was not taken into account in the Syrkin and Dyatkina calculations) provide serious objections to the resonance structure for B_2H_6.

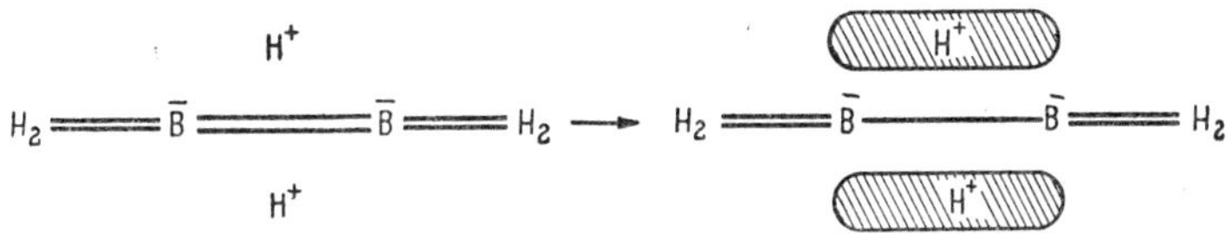

Figure 13.1. The 'protonated double bond' structure for diborane

(c) PITZER[8] has suggested an ethylene-like structure for diborane, each boron atom taking a formal negative charge, and the two protons being embedded in the π electron charge clouds of the double bond, forming a 'protonated double bond', *(see Figure 13.1)*. This interpretation appears quite attractive at first, particularly as the u.v. absorption spectrum is ethylenic in character, but it has been subjected to two main criticisms. First, the boron–boron distance is known to be 1·77 Å, which might seem rather too great for appreciable overlap of the $2p$ atomic orbitals of the negative boron atoms. Mulliken has calculated, however, that a considerable overlap does occur even at such a distance. The second criticism of Pitzer's model is that the two protons are relatively unscreened, and consequently might be expected to be acidic. Such acidic properties are not observed in B_2H_6.

(d) LONGUET-HIGGINS[1] has pointed out that the orbital functions used by Pitzer can be rearranged to give an alternative description of diborane in which the two BH_2 groups are linked by two equi-

valent 'three-centre' or 'banana' bonds, each of which embraces the two boron atoms and one of the bridge hydrogens. These equivalent orbitals resemble those discussed earlier for ethylene (page 108), except that each orbital now covers three atoms rather than two.

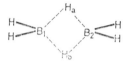

The functions used by Pitzer may be summarized as follows:

$\sigma_1 = sp^2$ orbital used by B_1 for σ bonding
$\sigma_2 = sp^2$ orbital used by B_2 for σ bonding
$\pi_1 = p_z$ orbital used by B_1 for π bonding
$\pi_2 = p_z$ orbital used by B_2 for π bonding
$S_a = s$ orbital of bridging atom H_a
$S_b = s$ orbital of bridging atom H_b

Two linear combinations of these atomic orbitals may be taken as follows:

$$\psi_1 = \mathcal{N}[\sigma_1 + \sigma_2 + \pi_1 + \pi_2 + 2S_a]$$
$$\psi_2 = \mathcal{N}[\sigma_1 + \sigma_2 - \pi_1 - \pi_2 + 2S_b]$$

$\mathcal{N}$ is a normalizing constant. An orbital outline for ψ_1 is derived in *Figure 13.2*.

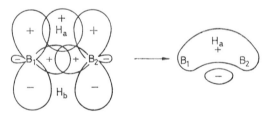

Figure 13.2. Orbital outline for a three-centre orbital

As *Figure 13.2* shows, the large banana-shaped positive lobe embraces H_a as well as the two boron atoms. The shape of the orbital described by function ψ_2 is directly analogous to that of ψ_1, except that the positive lobe now covers the bridging H_b. There are four electrons available for the bridge bonding, one from each boron and one from each hydrogen, so each of these molecular orbitals will be doubly filled.

At first sight the orbital outlines may be slightly misleading, since they may suggest that in the Pitzer description there is a concentration of charge between the boron nuclei (as a result of the σ bond), but that the charge concentration in the banana orbitals is away from the boron–boron axis. The difference is merely the result of an oversimplification of the orbital-outline diagrams; the charge distributions arise from identical wave-mechanical functions and must therefore be the same. The three-centre orbital description is, perhaps, to be preferred to the pro-tonated-double-bond one, since it may be considered as part of a series of polycentre orbitals.

We can get a simpler picture of the formation of the three-centre orbitals if we consider the boron atoms to be sp^3 hybridized—*cf.* the equivalent orbital treatment of ethylene on page 123. The simultaneous overlap of the $1s$ orbital of hydrogen by two sp^3 orbitals, one from each boron atom, gives a molecular orbital

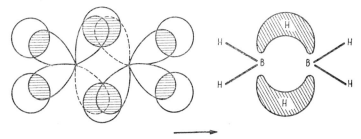

Figure 13.3. The 'three-centre orbital' structure for diborane
(The two orbitals which each boron atom uses in the bridge structure are equivalent, but since only one electron is available for the two orbitals, then one orbital is formally considered as vacant—a dotted outline—and the other singly occupied. A similar convention is used with subsequent diagrams for other bridge structures.)

covering all three atoms *(cf. Figure 13.3)*. This simplified deriva-tion is less rigorous, and really amounts to superposing the two con-tributing structures I and II of the resonance formulation (page 294), but it can be visualized more readily, especially when the more complicated electron-deficient higher hydrides of boron or metal alkyls are discussed.

THE STRUCTURES OF THE HIGHER HYDRIDES OF BORON
(*cf.* LIPSCOMB[9])

In recent years, considerable progress has been made in determining the structures of higher hydrides of boron such as B_4H_{10}, B_5H_9, B_5H_{11}, B_6H_{10}, and $B_{10}H_{14}$. The nuclear configurations of these

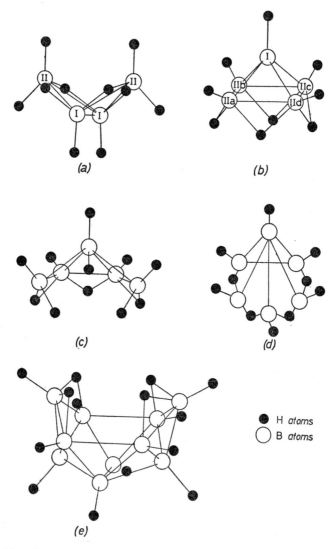

Figure 13.4. The nuclear configurations for the higher hydrides of boron

(a) B_4H_{10} (b) B_5H_9

(c) B_5H_{11} (d) B_6H_{10}

(e) $B_{10}H_{14}$

compounds have now been well established *(cf. Figure 13.4)*, although some finer details (B—H distances, for instance), are not known with certainty. Various experimental methods, including x-ray, micro-wave, electron diffraction, and nuclear magnetic resonance, have been used to obtain the structural parameters quoted in *Table 13.1*; where there is quite a number of similar B—B distances, the average value is quoted.

Table 13.1. Internuclear Distances in the Higher Boron Hydrides

Hydride	*B—B distance* (Å)			*Terminal B—H distance* (Å)	*Bridge B—H distance* (Å)	
B_4H_{10}	1·75 (1)*	1·84 (4)		1·19	1·33 (4)	1·43 (4)
B_5H_9	1·69 (4)	1·80 (4)		1·22	1·35	
B_5H_{11}	1·74 (5)	1·87 (2)		1·09	1·24	
B_6H_{10}	1·60 (1)	1·74 (6)	1·79 (3)	—		
$B_{10}H_{14}$	1·76 (19)	2·01 (2)		1·25	1·34 (4)	1·42 (4)

* The figures enclosed in parentheses represent the number of such bonds in the particular structure.

The bonding in these higher hydrides of boron can be described quite simply in terms of the overlap of the hydrogen $1s$ orbitals with either sp^3, sp^2 (and one $2p$), or sp (and two $2p$), orbitals of the boron atom. The terminal B—H bonds are considered as normal two-electron bonds, and the bridging B—H—B bonds are three-centre molecular orbitals. Any electrons remaining are used to bind the boron atoms in the framework. The boron–boron bonds are either normal two-centre two-electron bonds, or multi-centre bonds. With three-centre bonds, the orbital can either be banana-shaped as in diborane, but with a boron atom taking the place of the hydrogen, or it can be rather more like a 'steering wheel', being formed by the mutual overlap of three orbitals (one from each boron) which point towards the centre of the equilateral triangle formed by the three boron atoms *(cf. Figure 13.5)*.

These principles will now be used to discuss the valency structures of two of the individual hydrides.

1. B_4H_{10}—The boron atoms labelled B_{II} in *Figure 13.4 (a)* each form two terminal B—H and two bridge B—H—B bonds, so that all the four sp^3 orbitals of boron are accounted for. Each B_I atom, however, forms one terminal B—H bond and two B—H—B

bridge bonds, leaving each B_I atom with a single sp^3 orbital containing one electron. These singly-filled orbitals overlap to give a B_I—B_I bond. The bridge bonds are asymmetrical because of the non-equivalence of the boron orbitals (*i.e.* the two orbitals used for the terminal B—H bonds differ slightly from those used for the bridge B—H—B bonds), and the unfavourable angles ($HB_{II}H$) between the bridge bonds.

2. B_5H_9 *(Figure 13.4 (b))*—The four B_{II} atoms each form one terminal B—H and two bridge B—H—B bonds, and therefore have one orbital and one electron to contribute to the framework. The single B_I atom, however, forms only one terminal B—H bond,

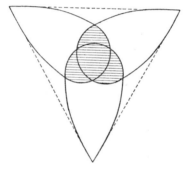

Figure 13.5. The formation of a central 3-centre orbital

leaving three orbitals and two electrons for the framework. For convenience, these B_I orbitals can be considered as a single sp orbital (the other is used for the terminal B—H bond)—we shall take its axis as the x axis—and two unchanged p orbitals. Three bonding molecular orbitals can be formed from these atomic orbitals:

(i) By combining the B_I sp orbital with orbitals from each of the four B_{II} atoms, giving a molecular orbital covering all five boron atoms, *i.e.* a five-centre orbital,

$$\psi = \frac{1}{\sqrt{2}}\psi_{Isp} + \frac{1}{2\sqrt{2}}(\psi_{IIa} + \psi_{IIb} + \psi_{IIc} + \psi_{IId})$$

(ii) By combining the B_I p_y orbital with orbitals from the B_{IIa} and B_{IIc} atoms, giving a 'banana' molecular orbital covering three boron atoms

$$\psi = \frac{1}{\sqrt{2}}\psi_{Ipy} + \frac{1}{2}[\psi_{IIa} - \psi_{IIc}]$$

(iii) By a similar combination of the $B_I p_z$ orbital with orbitals from the B_{IIb} and B_{IId} atoms, giving another 'banana' bond

$$\psi = \frac{1}{\sqrt{2}}\psi_{Ipz} + \frac{1}{2}[\psi_{IIb} - \psi_{IId}]$$

Since there are six electrons to feed into these bonding molecular orbitals, each orbital is completely filled.

There are, of course, quite a number of other boron hydrides besides those we have discussed, and in some cases their structures (and those of related substitution products such as $B_{10}H_{12}(CN)_2$) have been investigated. By using the basic concept of three-centre orbitals, Lipscomb has developed a topological approach to predict which hydrides are capable of existence, and interested readers are referred to his review article for an outline of the procedure, as well as for a discussion of bonding in hydrides other than the ones we have commented on.

THE STRUCTURES OF THE BOROHYDRIDES

When diborane reacts with organo-metallic compounds, borohydrides, of the general formula $M(BH_4)_n$, are formed, which often have the electron-deficient characteristics of diborane itself.

There is an interesting gradation in the physical and chemical properties of the borohydrides of lithium, beryllium, and aluminium. *Table 13.2* gives the melting and boiling points of these substances. A significant difference is also found in their reactivity

Table 13.2

	m.p. (°C)	b.p. (°C)
$LiBH_4$	275 (*decomposition*)	—
$Be(BH_4)_2$	123 (*under pressure*)	91·3 (*sublimes*)
$Al(BH_4)_3$	−64·5	4·5

towards typical electron-donor molecules such as trimethylamine; lithium borohydride does not react at all, unlike both beryllium and aluminium borohydrides which react readily even at low temperatures. It is, thus, apparent that lithium borohydride is not electron-deficient, being undoubtedly ionic, containing Li^+ cations and tetrahedral $[BH_4]^-$ anions. Similar ionic structures can be put forward for the analogous lithium aluminium hydride, $Li^+[AlH_4]^-$, and lithium gallium hydride, $Li^+[GaH_4]^-$.

The borohydrides of beryllium and aluminium, on the other hand, are predominantly covalent, as shown by their high volatility, and electron-acceptors, as indicated by their reaction with trimethylamine. Electron-diffraction experiments indicate bridge structures analogous to that of diborane

Since the hydrogen atoms bridge atoms which are not identical (*i.e.* boron and aluminium, or boron and beryllium), it is not surprising to find that the bridges are unsymmetrical. The electrons in these bridged borohydride structures are presumably allocated much as in diborane itself, and the bonding is therefore best described in terms of three-centre molecular orbitals. By forming such a bridge system, both beryllium and aluminium achieve considerable symmetry; beryllium thus assumes a tetrahedral, and aluminium an octahedral, bond distribution.

THE METAL ALKYLS (*cf.* COATES[10])

Alkyls formed by elements of the first three periodic groups (alkali metals, alkaline earth metals, and the boron group), are particularly interesting from a valency point of view because many of them are polymeric. We shall see that the forces holding the units together are similar to those in diborane. The detailed structures of a number of alkyls have been evaluated by x-ray methods, and significant physical data is available for many of the others; this may be summarized as follows:

Alkali metals—As we saw in Chapter 11 the solid lithium methyl and ethyl contain tetrameric units, and in the case of the former the methyl groups are in the centre of the faces of the Li_4 tetrahedra. Other lithium alkys are polymeric in solution but there is no further structural information.

Alkaline earth metals—Solid beryllium dimethyl is a polymeric substance, and contains chains of beryllium atoms bridged by methyl groups. The Be—C—Be angle is 66°, and each Be atom is surrounded tetrahedrally by four Me groups

The vapour of beryllium dimethyl is a rather complex equilibrium mixture (COATES[11] et al.) containing mainly dimer, with small amounts of trimer, and monomer molecules. The structure of the monomer presents no difficulties; the beryllium atom makes use of sp hybrid orbitals giving a linear structure, CH_3—Be—CH_3, while the dimer and trimer molecules presumably have the structures

$$CH_3 \underset{}{\overset{}{—}} Be \underset{CH_3}{\overset{CH_3}{\diagdown\diagup}} Be \underset{}{\overset{}{—}} CH_3 \qquad CH_3 \underset{}{\overset{}{—}} Be \underset{CH_3}{\overset{CH_3}{\diagdown\diagup}} Be \underset{CH_3}{\overset{CH_3}{\diagdown\diagup}} Be \underset{}{\overset{}{—}} CH_3$$

containing methyl bridges.

The other beryllium alkyls are also polymerized but to a smaller extent, and this shows up in their physical properties; thus whereas the dimethyl is a solid which sublimes at 217°C (760 mm), the di-ethyl and di-isopropyl are both viscous liquids.

Magnesium dimethyl has also been studied recently (WEISS[12]) and found to have an approximately tetrahedral arrangement of methyl groups about magnesium, with two methyls between each pair of metal atoms.

Boron Group

These elements form an interesting series of alkyls (MULLER and OTERMAT[13]) because whereas the boron compounds are monomeric, the aluminium ones are dimeric, and those of the heavier elements monomeric. A single-crystal x-ray diffraction study of aluminium trimethyl has shown the molecule to be a bridged structure. Electron diffraction experiments do not distinguish clearly between a bridged and an ethane-like structure, but on the whole the bridge structure is favoured.

$$CH_3 \underset{CH_3}{\overset{CH_3}{\diagdown}} Al \underset{\underset{H_3}{C}}{\overset{\overset{H_3}{C}}{\diagdown\diagup}} Al \underset{CH_3}{\overset{CH_3}{\diagup}}$$

The four terminal methyl groups and the two Al atoms are co-planar; the two bridging methyl groups are symmetrically placed above and below this plane equidistant from the two aluminium atoms. The Al—C—Al angle in the bridge is about 70°, a significantly low value; the terminal C—Al distances are 1·99 Å and the bridge C—Al distances are 2·24 Å.

Platinum

An x-ray study of so-called platinum tetramethyl indicated that it was tetrameric with a slightly distorted cubic arrangement in which a methyl group bridged three platinum atoms. Attempts[14] to repeat this work have been unsuccessful, and it is probable that the compound originally studied was $Pt(OH)Me_3$, with OH groups forming the bridges.

It will be noticed that all the metal methyls have a structural feature in common, namely, the bridging of the metal atoms by a methyl group, and this gives structures with at least a formal resemblance to diborane, with methyl groups now replacing the bridge hydrogens. The chemical bonds in these alkyls are evidently of the same type; the bonds in aluminium trimethyl will be discussed at some length, but the arguments invoked may be directly applied to the other molecule as well. The bonding is necessarily electron-deficient, and any interpretation should account for the small M—C—M angle in the bridge.

In aluminium trimethyl, as in diborane, the terminal bonds (Al—CH_3 in this case) are normal, and the electron-deficiency is confined to the bridge bonds where there are only four electrons available to provide the four Al—C bonds. Explanations put forward for these Al—C bonds must also explain how the carbon atoms in the bridge form five covalent bonds.

The main interpretations of the bonding closely follow those of diborane, and suggestions that linkages might involve hydrogen bonding, or resonance (between forms such as I to IV) have been made.

$$I \qquad\qquad II$$

$$III \qquad\qquad IV$$

By analogy with the protonated double bond description of diborane, a 'methylated double bond' structure has been suggested for $Al_2(CH_3)_6$, in which the $\overset{+}{C}H_3$ groups are embedded in the π electron charge clouds of an Al—Al double bond.

$$\overset{+}{C}H_3$$

$$\overset{-}{Me_2Al}=\overset{-}{AlMe_2}$$

$$\overset{+}{C}H_3$$

Probably the most satisfactory explanation for the bridge bond in these alkyl structures uses central 3-centre orbitals. The molecular orbitals may be compounded from the sp^3 hybrid orbitals of aluminium and carbon; each molecular orbital covers two aluminium and one carbon atom, and is completely filled *(see Figure 13.6 (a))*. If there is to be maximum overlap of the carbon sp^3 hybrid orbital by the two aluminium sp^3 orbitals, then the

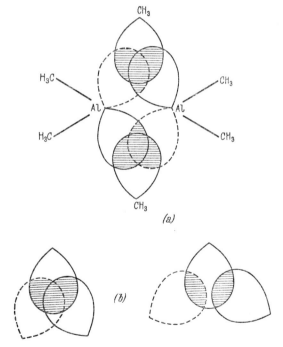

Figure 13.6. (a) The formation of 'three-centre orbitals' in $Al_2(CH_3)_6$
(b) The overlap of a carbon sp^3 orbital by two aluminium sp^3 orbitals : the acute-angled structure gives greater overlap, and hence a stronger bond.

Al—C—Al ngle must be small, as emphasized in *Figure 13.6 (b)*. This bonding description also accounts for the lower degree of polymerization found with the higher alkyls, since the heavier alkyl groups would be much less effective in bridging the metal atoms. MULLER and OTERMAT have discussed in some detail the factors important in determining whether or not alkyl bridging takes place, with particular reference to the boron group alkyls. They point out that for bridging there should be (i) a large electro-negative difference (M and C), (ii) a low $s \rightarrow p$ promotion energy for M, (iii) a large M—C bond energy, and (iv) minimum inner shell repulsions between the two metal atoms. The first two factors are unfavourable for boron and the latter two unfavourable for the heavier elements; thus it is only with aluminium trimethyl that dimerization occurs.

A comparison of the hydrides of boron, aluminium, and gallium with the methyls shows that the bridged hydrides are more poly-merized than the alkyls. Now an s orbital can be overlapped by two other orbitals more readily than can an sp^3 hybrid orbital, and the amount of overlap is less dependent on the bond angle, so that hydrogen atoms are more suitable for bridging than methyl groups, and hydrides will accordingly be more polymeric than alkyls.

REFERENCES

[1] LONGUET-HIGGINS, H. C. *Q. Rev. chem. Soc.* 11 (1957) 121.
(This review gives structural information available up to this date, and should be consulted for further references.)
[2] SMITH, H. W. and LIPSCOMB, W. N. *J. chem. Phys.* 43 (1965) 1060
[3] BARTELL, L. S. and CARROLL, B. L. *J. chem. Phys.* 42 (1965) 1135
[4] STITT, F. *J. chem. Phys.* 9 (1941) 780
[5] SHOOLERY, J. N. *Discuss. Faraday Soc.* 19 (1955) 215
[6] BURAWOY, A. *Nature, Lond.* 155 (1945) 328
[7] SYRKIN, Ya. K. and DYATKINA, M. E. *Acta phys.-chim. URSS* 14 (1941) 547
[8] PITZER, K. S. *J. Am. chem. Soc.* 67 (1945) 1126
[9] LIPSCOMB, W. N. *Adv. inorg. Chem. Radiochem.* 1 (1959) 118
[10] COATES, G. E. *Organometallic Compounds.* 2nd edn. 1960 Methuen, London,
[11] COATES, G. E., GLOCKLING, F. and HUCK, N. D. *J. chem. Soc.* (1952) 4496
[12] WEISS, E. *J. organomet. Chem.* 2 (1964) 314
[13] MULLER, N. and OTERMAT, A. L. *Inorg. Chem.* 4 (1965) 296
[14] BRUBAKER, C. H. *Private communication*

AUTHOR INDEX

Page numbers in brackets refer to references

308

SUBJECT INDEX

The more important page numbers are printed in heavy type.

314